AGRICULTURAL REFORMS AND
GRAIN PRODUCTION IN CHINA

STUDIES ON THE CHINESE ECONOMY

General Editors: Peter Nolan, Lecturer in Economics and Politics, University of Cambridge, and Fellow and Director of Studies in Economics, Jesus College, Cambridge, England; and Dong Fureng, Professor, Chinese Academy of Social Sciences, Beijing, China

This series analyses issues in China's current economic development, and sheds light upon that process by examining China's economic history. It contains a wide range of books on the Chinese economy past and present, and includes not only studies written by leading Western authorities, but also translations of the most important works on the Chinese economy produced within China. It intends to make a major contribution towards understanding this immensely important part of the world economy.

Published titles include:

Dong Fureng
INDUSTRIALIZATION AND CHINA'S RURAL
  MODERNIZATION

Du Runsheng (*edited by Thomas R. Gottschang*)
REFORM AND DEVELOPMENT IN RURAL CHINA

Qimiao Fan and Peter Nolan (*editors*)
CHINA'S ECONOMIC REFORMS

Christopher Findlay, Andrew Watson and Harry X. Wu (*editors*)
RURAL ENTERPRISES IN CHINA

Gao Shangquan
CHINA'S ECONOMIC REFORM

Michael Korzec
LABOUR AND THE FAILURE OF REFORM IN CHINA

Nicholas K. Menzies
FOREST AND LAND MANAGEMENT IN IMPERIAL CHINA

Ryōshin Minami
THE ECONOMIC DEVELOPMENT OF CHINA

Agricultural Reforms and Grain Production in China

Shujie Yao
Senior Lecturer in Economics
University of Portsmouth

St. Martin's Press

First published in Great Britain 1994 by
THE MACMILLAN PRESS LTD
Houndmills, Basingstoke, Hampshire RG21 2XS
and London
Companies and representatives
throughout the world

A catalogue record for this book is available
from the British Library.

ISBN 978-1-349-23555-1   ISBN 978-1-349-23553-7 (eBook)
DOI 10.1007/978-1-349-23553-7

First published in the United States of America 1994 by
Scholarly and Reference Division,
ST. MARTIN'S PRESS, INC.,
175 Fifth Avenue,
New York, N.Y. 10010

Library of Congress Cataloging-in-Publication Data
Yao, Shujie.
Agricultural reforms and grain production in China / Shujie Yao.
p.   cm. — (Studies on the Chinese economy)
Includes bibliographical references and index.
ISBN 0–312–12370–1
1. Agriculture—Economic aspects—China.   2. Grain—China.
I. Title. II. Series.
HD2097.Y38   1994
338.1'851—dc20   94–21687
                                             CIP

To my beloved grandmother Zheng Sujin

Contents

vii

List of Figures

List of Tables

Acknowledgements

I am most indebted to Dr Ajit Bhalla of the International Labour Organization for undertaking the seemingly endless task of reviewing the entire book. His continuous encouragement and valuable advice on various issues kept me on a straight and narrow path in its preparation. I am also grateful to him for allowing access to all his recent research work on India and China, which provided a lot of highly relevant and informative knowledge. Dr Peter Nolan, of Jesus College, Cambridge, deserves my particular gratitude for his advice on structural and presentational issues. My appreciation is also offered to Mr William van der Geest of Oxford University for his valuable comments on Chapter 3. Also at Oxford I would like to thank Professor George Peters, Dr Frances Stewart, Dr Roger Hay and Mr Alex Duncan for their comments on Chapters 4 and 6. Among my immediate colleagues at Portsmouth University, the very supportive attitude of Mike Dunn and Guy Judge is gratefully acknowledged. They arranged for the entire appendix to be typed at the Portsmouth Business School. My thanks are also extended to my former supervisors, Professor David Colman, Dr Trevor Young and Dr Noel Russell, at Manchester University, for their guidance on my earlier research work which laid the foundation for this book. In China, many experts and friends of the Ministry of Agriculture deserve my thanks. Among these are Professor Guo Shutian, Messrs Yang Qirong, Liu Jiemin, Xia Jiangmin and Qin Rei who gave me the most recent data and their research on agriculture and rural township and village enterprises (RTVEs). Needless to say none of those mentioned is in any way responsible for any errors that I may have committed. I accept full responsibility for those.

Finally, I would like tó thank Miss Jane Marchant, at the Department of Accounting and Business Information Systems of Portsmouth University, for her heroic efforts in typing and processing the four chapters of the Appendix. Also at Portsmouth Business School, a special debt of thanks is owed to Mrs Val McDermott for her generosity and enormous efforts in editing the final drafts of various chapters. She did this with unfailing grace and humour. However, much of the strain was transmitted to my family who had to tolerate my overtime working at home or in the office for a countless number of days. Xiaowen, Joseph and Julia deserve my special thanks and love.

SHUJIE YAO

Preface

Sustainable growth in agricultural and grain production is important not only to raise the living standards of the population; it is also a prerequisite for rapid economic growth in China. The development experience in pre-reform China coincided with the economic thought of the 1950s when emphasis in development policy was placed on urban industrial growth, with agriculture being treated as a source of inputs for the manufacturing sector. The economic structure resembled that described by Lewis (1954). In the literature, it was Kuznets (1964), Mellor (1976) and others who pointed out the positive role of agriculture as an engine of development and emphasized the importance of 'balanced growth' between agriculture and industry. Agricultural reforms in China and elsewhere in the world since the late 1970s have confirmed the need for a 'balanced growth' strategy in the development process.

As grain is the single most important enterprise in Chinese agriculture and the focus of many important policy issues, this book is concentrated on the problems of grain production. The overall objective is to prove and reinforce the need for smooth and sustainable growth in agriculture and grain production in the Chinese economy by comparing the performance of the national economy in general, and grain and agricultural production in particular, before and after economic reforms. The turning point came in 1978 when the country started implementing a series of reform policies in the rural areas. So far agricultural reforms have succeeded in many aspects. Living standards of the rural population have been substantially raised. Absolute poverty in rural areas has been largely eliminated. Production of many key agricultural products, including cereals, fish, meat, cotton and oil-crops, has climbed up many record-breaking steps. The production structure both inside and outside agriculture has been significantly rationalized, enabling resources to be used far more efficiently.

The most notable structural change is the shift of traditional farming towards non-farm enterprises in the rural areas. The so-called RTVEs have emerged to become the most dynamic sector of the national economy. In only fifteen years of development, the total output value of RTVEs has outstripped that of the entire agricultural sector. At the time of writing, it accounts for over one-third of the national industrial output and one-quarter of the total national exports. Over 20 per cent of the rural labour force are engaged in non-farm production which has become the most important source of income of farmers in many areas.

Rural and agricultural reforms over the last fifteen years have also had many problems. The performance of agriculture and grain production has been by no means smooth. The rapid development of rural industries has also created many challenges and problems. However, not all of these problems can be attributed to the reform itself as maintained by a number of economists in China and abroad. The major problems come from the methods of implementation, the sequencing and timing of reforms, incompetent government macro-economic policies and, above all, the lack of will of decision makers, both at the central and local levels, to put agriculture in its appropriate place at the centre of the economy.

To solve China's agricultural problems, it is important to put agriculture into the national economic context. Sicular (1993) argues that many of the agricultural problems in China are caused by two macro-economic factors: the state budget and urban bias. We share exactly the same view. The earlier stage of agricultural reforms (1978–85) was characterized by institutional reforms as well as price incentives. Raising agricultural prices without correspondingly reducing urban subsidies forced the government into persistent budgetary deficits. Once agriculture and grain production reached their unexpected historic high in 1984, it was not surprising that the government, with an apparent urban bias, reduced the budgetary deficit by subduing agricultural prices. The policy changes in 1985 were a typical example of the lack of government commitment to agriculture and the rural population. The phenomenon of widespread IOUs issued to farmers for their deliveries of grain to the state in the early 1990s is simply a further illustration of the bias. During a short span of fifteen years of economic reforms, China has experienced at least three distinctive cycles of booms and busts. During the boom years, large investment funds are channelled to finance industrial and urban construction projects with few monies left for agriculture. In the difficult periods, agriculture is the first to be targeted for budgetary restraint.

In recent years, most western countries have been impressed by the rapid growth of the Chinese economy, but China is in fact facing a number of dangerous problems. These include the uncontrollable expansion of industrial and housing construction; highly skewed income distribution between rich and poor, and between urban and rural areas; high inflation; regional disparities and corruption. Some of these problems are seemingly unrelated to agriculture but they will affect agriculture negatively. On the other hand, some of these problems can be easily resolved if the government can sincerely put agriculture first as it always states in its policy documents.

Another way to solve the agricultural problems is to increase internal efficiency within agriculture. This involves further rationalization of the production structure. One important aspect of agricultural reform in China is the shift of production from low-value enterprises such as grains to high-value ones such as cash crops and livestock products. Another area of structural adjustment involves the exploitation of regional comparative advantages in production. Chinese agriculture has made tremendous progress in these areas. Indeed, raising internal efficiency in agriculture and grain production is one of the major topics covered in this book. In post-reform China, rationalization of agricultural production structure has been facilitated by government policies in two aspects. Firstly, free marketing of agricultural products is allowed and government mandatory control on internal and external trade is greatly reduced. Secondly, relative prices of commodities and spatial price differentials reflecting resource scarcity and regional comparative advantages have been formulated under the reform. In this book, however, we argue that there still exists a huge potential of internal efficiency to be exploited inside the agricultural sector.

To study the ways of exploiting regional comparative advantage in production, qualitative description is not sufficient. There is also a need for quantitative analysis. Quantitative analysis can take many approaches. The quantitative approach employed in this book is that of a quadratic spatial equilibrium model to examine the costs and benefits of various government policy scenarios. The commodity under scrutiny is grain which covers rice, wheat, corn and other cereals as an aggregate product. The advantage of this approach is that it can simultaneously identify the optimal spatial patterns of supplies, demands and prices under the specified policy regime. It can also calculate the exact costs and benefits for individual groups of people in individual regions as well as the net loss or gain to society induced by a specific policy. The disadvantages of this approach are that it requires data which are not easily obtainable, and that it requires much mathematical and computational knowledge. Thus some researchers and readers may find it too onerous to appreciate. It is because of this later disadvantage that the modelling work has been placed in a self-contained appendix. This becomes a distinctive feature of the book.

The book is divided into three parts. Part I contains four chapters which describe the performance of agriculture and grain production before and after the economic reforms. It identifies the major problems under the old policy regime before 1978 and discusses the reasons for success and the new problems arising after the reforms. Ample data and empirical observations are provided to support our arguments and hypotheses. Part II, containing Chapters 5 to 7, is more policy-oriented. It demonstrates the diminishing role of agriculture in the national economy in terms of its contribution to GDP. We support the view that it is necessary and inevitable to have a faster growth in the non-farm sectors than in the farming sector. We also maintain that a balanced growth between agriculture and industry is an essential requirement in the development process of the Chinese economy. Part III contains the appendix material. It includes four technical chapters which aim to quantify the effects of a number of important policy scenarios. The mathematical models are complicated and difficult to understand. However, we have tried our best to make the algebra and notations as readable as possible. It is expected that most readers would find it much easier than it looks by going through all the chapters of the appendix. For those who do not appreciate such mathematical rigour, they can just read Chapter 7 which summarizes the computational results of the models.

Part I
Agricultural Policies and Economic Reforms

Introduction

This part of the book reviews Chinese agricultural policies and their effects on economic growth, especially grain production in different periods since independence in 1949. We argue that economic performance and grain production were greatly influenced by state policies on rural and agricultural development. This discussion also provides an economic background for the econometric analyses in the later parts of the book concerning the quantification of the effects of some major agricultural policies on the patterns of grain production across regions.

Although China is a vast country, with a total area of more than 9.6 million square kilometres, only 10 per cent of this huge territory can be cultivated. In other words, China has only 7 per cent of the world's total arable land, but it has to feed more than 20 per cent of the world's total population. Since 1949 agricultural land has been declining due to industrialization and urban development. For example, the total area under farm crops decreased by 9.2 per cent between 1956 and 1989. The area sown with grain crops has been declining even faster. It decreased from 136.3 million hectares in 1956 to 112.2 million hectares in 1989, or by almost 18 per cent in 33 years (see Chapter 4, Table 4.4). On the other hand, the population has been growing rapidly. It increased from 575 million in 1952 to 1127 million in 1989 (Chapter 2, Table 2.11), or by 96 per cent in 37 years, with an annual rate of increase of 1.78 per cent.

Increasing population and decreasing arable land are two major constraints on agricultural development in China. Since China is still one of the poorest countries in the world, agriculture (especially grain production) is of vital importance in feeding such a huge and expanding population. It is also important as a resource base for industrialization. The importance of grain self-sufficiency has always been noted by the politicians and by many economists at home and abroad. China cannot rely on large-scale importation of foodgrain. Even if China could afford more imports to supplement its own food production, the scale of its food needs could put severe strains on the world market should China decide to buy from abroad as little as 10 per cent of its food requirement.

However, food self-sufficiency is not the only objective of government agricultural policy in modern China. The desire to industrialize has been another goal of the government. In fact, throughout the post-independence period, agriculture in general, and food production in particular, have been used as the primary source of capital accumulation for industrial development. The conflict between industrialization and agricultural development has been the most distinctive dilemma in the Chinese agricultural policies. Economic performance and agricultural development have been largely dependent on the state's ability to balance these two objectives.

In addition, the post-independence period has been characterized by many political and ideological changes. Some of these changes have had significant implications for agricultural development and grain production but their motives are not easy to understand. This makes policy analysis particularly difficult. In this book, therefore, discussion is not limited to economic analysis. It will also be extended to cover political events and their effects on economic performance.

Agricultural policies and political events have been highly inconsistent over time, but as regards the degree of state intervention in agricultural production, agricultural policies can be divided into two different types. The first type of policy is characterized by less state control in agriculture and a more balanced development strategy between agriculture and industry. As illustrated in Table PI, this type of policy is observed in the following sub-periods: the period covering land reform and the First-Five-Year Plan (1949–57), the recovery period in the early 1960s (1962–65), and the recent economic reform (since 1978 onwards). During all these sub-periods, both agriculture and industry performed much better and the living standards of the people, especially the farmers, increased much faster than in the other sub-periods.

The second type of policy is characterized by strict state control of agriculture and a strong bias towards industry at the expense of agriculture. This is observed during the Great Leap Forward movement (1958–61) and the Cultural Revolution (1966–78). The Great Leap Forward movement evolved into the most devastating famine in China and led to a total collapse of the national economy. The Cultural Revolution resulted in the longest period of stagnation in agriculture and farmers' living standards.

Part I consists of four chapters. Chapter 1 discusses the performance of agriculture and grain production for the first two periods outlined in Table PI. It explains why rapid growth in national income and

Table PI Political events and agricultural policies

Periods	Political events	Major agricultural policies
1949–52	Socialist land reform	(1) Specialized production (2) Free market allowed
1953–57	Co-operative movement	
1958–62	Great Leap Forward movement	(1) Compulsory commandism for demand and supply (2) No free markets (3) Grain self-sufficiency for all localities
1963–65	Readjustment period	(4) Very low prices for agricultural products (5) Commune system
1966–78	Cultural Revolution	(6) Policies during 1963–65 similar to those in the recent reforms
1979–	Agricultural reforms	(1) Relatively higher prices for agricultural produce (2) Free markets encouraged (3) Specialized production encouraged (4) Abolition of the commune (5) Various kinds of production responsibility, especially household contract responsibility

Source: Yao and Colman (1990).

industrial production did not increase the living standards of the people and why rapid changes in production structure were not accompanied by corresponding changes in the structure of employment before 1978. Three policies are identified as responsible for these paradoxes: (1) overemphasis on the development of heavy industry; (2) overemphasis on local grain self-sufficiency and state control on marketing; and (3) forced collectivization, especially the institution of the commune system.

In contrast with the performance in the pre-reform era, economic reforms in the 1980s have resulted in rapid agricultural growth and a vast improvement in the living standards of the whole rural population. Per capita income and food production have increased steadily; the share of labour engaged in the traditional agricultural sector, especially crop production, has been substantially reduced; the production of non-food crops has experienced even greater growth; rural industries have become the major production and income generating activities in most fast-growing regions, especially the coastal areas.

Economic reforms since 1978 have brought about two sets of distinctive changes. The first includes high growth rates of national income and significant improvement in the living standards of the entire population, especially the rural population. The second includes rapid growth in production and dramatic changes in the structures of production and employment. These changes are described in Chapter 2. To help the readers understand the extent of changes in rural China during the reforms, my personal observations on the changes of production and living standards in a remote village in eastern Guangdong are presented in this chapter. The village can be regarded as an average village in eastern Guangdong. Thus the changes described are representative for that part of the province. Guangdong is one of the most prosperous regions in China and has experienced the most successful reforms in the last decade. However, its reform experiences are not representative of the whole nation.

Dramatic changes in China due to economic reforms have drawn more and more attention at home and abroad. The reforms implemented so far are considered to be very successful by many Western and Chinese economists. There are, however, various explanations for this success. Some authors, such as Lardy (1983), Griffin (1984), Riskin (1987), Kojima (1985) and Johnson (1988), have pointed out that the successful performance of the new agricultural policies is due to more state investment, more incentives for the peasants to work and more emphasis on the specialization of production. Other authors, like Raj (1983) and Ghose (1984), have argued that the success is due mainly to price factors incorporated with the infrastructure laid down during the Great Leap Forward and the Cultural Revolution. Ghose went further, concluding that Chinese agricultural reforms were actually maximizing the short-term benefit at the expense of the state budgetary costs which would in turn retard long-term development. As shown (in Tables 2.10, 2.11, 2.12) in Chapter 2, the relatively poor performance between 1985 and 1988 compared with the earlier years after the reforms, especially in the production of grain and cotton, seems to support such an argument. However, the later recovery of grain production in the 1989–91 period suggests that this conclusion is tentative. The reasons for unstable performance of grain production and agriculture as a whole are much more complicated.

Chapter 3 discusses the policies responsible for the success in economic growth and agricultural production during the reforms. We conclude that: (1) the agricultural production responsibility system, especially the household contract system, is the key to success in rural reforms; the benefit of any economies of scale arising from the commune system is overshadowed by individual incentives arising from the responsibility system; (2) specialization in production and market liberalization characterized by price reform and freer trade are two principal factors accounting for increased allocative efficiency and farmers' incomes; (3) the abolition of the commune system has substantially reduced the overheads of agricultural production and facilitated sustained agricultural and rural development; and (4) although the infrastructure built during the previous two decades certainly helped the sudden upsurge of production in the early years of the post-Mao reforms, sustained agricultural growth was possible mainly because of the introduction of economic reforms and increased incentives to farmers; (5) a setback of grain and agricultural production during 1985–88 after the record high performance in 1984 was due to a number of factors but the lack of state investment in agriculture and reduced price incentives were important among these.

The prospects of grain production under economic reforms are discussed in Chapter 4. It analyses the reasons for the stagnation of agricultural and grain production between 1985 and 1988 and the recovery up to 1991. The major constraints on future agricultural and grain production in China are still the decreasing area of arable land and the decreasing relative profitability of grain production. However, the need to achieve long term food self-sufficiency cannot be underestimated. This is justified by many factors: (1) China still has a comparative advantage in food production; (2) the need for stability of supply in domestic market; and (3) the need to save foreign exchange. The potential for raising grain production lies in the ability to increase land productivity through greater inputs of technology and materials. It also depends on the ability of the state to provide necessary services, and particularly on its commitment to agricultural investments.

1 Agricultural Policies before Reforms

1.1 INTRODUCTION

Economic development between 1949 and 1978 can be divided into two different sub-periods. The first sub-period includes land reform and the First-Five-Year Plan of 1949–1957. Industrial and agricultural production increased rapidly and people's living standards were greatly improved. The second sub-period includes the Great Leap Forward movement and the Cultural Revolution. The Great Leap Forward movement (1958–62) was characterized by the 'Three Red Flags': the People's Commune, Agriculture Learning from Dazhai and Industry Learning from Daqiang. The movement resulted in the most devastating famine in Chinese history. The so-called 'Three Difficult Years' of 1959–61 saw millions of people starve to death. All economic indicators had significantly negative growth rates. The readjustment period of 1963–5 helped to restore the economy but it was followed by the Cultural Revolution from 1966 to 1978. The Cultural Revolution resulted in the longest political turmoil and economic stagnation since 1949.

Despite many disastrous political events, China achieved impressive economic and agricultural growth during the pre-reform period. Although there were marked fluctuations in growth over time, especially during the Great Leap Forward movement and the Cultural Revolution, all major economic indicators, including per capita national income, per capita cereal output, and per capita industrial output, had positive and significant growth rates over the whole pre-reform period. National income in real terms quadrupled from 1953 to 1978, with an annual growth rate of 5.68 per cent which was 3.68 per cent higher than population growth (see Chapter 2, Table 2.1). The structure of the national economy also experienced significant changes. The share of agriculture in national income decreased whereas that of industry increased rapidly. During 1953–78, the share of agriculture in national income decreased by almost 2 per cent per year but that of industry increased by more than 3 per cent per year.

8

Rapid economic growth and structural changes, however, did not significantly improve the living standards of farmers. Empirical observations suggested that the living standards of the Chinese farmers suffered from marked fluctuations during the Great Leap Forward movement and the Cultural Revolution. Real per capita income stagnated or increased marginally between 1953 and 1978. Structural changes in the national economy did not greatly affect the structure of employment and population. By 1978, 80 per cent of the total population still lived in the rural areas and engaged in traditional farming.

There are a number of paradoxical features in the Chinese agricultural sector of the pre-reform period. The first is the change in structure of the national economy and the composition of population and employment. Although the share of agriculture in the economy decreased considerably between 1952 and 1978, the share of rural population remained almost unchanged over the same period. A second paradox is the persistence, between the mid-1950s and the mid- to late 1970s, of chronic malnutrition and low income in a significant proportion of the rural population despite a doubling of per capita national income between these two periods. The last paradox is Mao Zedong's ideology of socialist collectivization, especially the commune system. While he led a rural-based Party to power on the promise of building an economically advanced socialist country and improving the welfare of the peasantry, in retrospect many of the policies pursued by the Party after 1949 appeared to have fundamentally undervalued agriculture and left a large proportion of Chinese rural population enmeshed in poverty for more than two decades.

In this chapter, we attempt to resolve these paradoxes by drawing on three different explanations: (1) the allocative efficiency of state investment; (2) the pursuit of local food self-sufficiency and state intervention in agricultural marketing; and (3) institutional change. The bias of state development policy towards heavy industry was the main reason for the slow change in labour structure and low investment efficiency. The pursuit of local food self-sufficiency suppressed the production of non-food agricultural commodities and reduced allocative efficiency within agriculture because it ignored regional comparative advantages in production. Institutional change, especially collectivization, suppressed labour incentives and artificially increased non-productive overheads incurred by excessive numbers of commune and production team cadres.

1.2 AGRICULTURAL PERFORMANCE

Economic development in China accelerated rapidly after the civil war. By 1952 the economy had almost recovered the pre-war peak level. Real national income more than quadrupled between 1953 and 1978 (Chapter 2, Table 2.1), registering an annual growth rate of 5.68 per cent. This was in marked contrast to the first half of the twentieth century, when there was no sustained per capita growth (Perkins, 1975; Lardy, 1983).

Dramatic growth was accompanied by far-reaching changes in the structure of the national economy. The share of national income originating in industry rose from 20 per cent in 1952 to nearly 50 per cent in 1978, whilst agriculture's share shrank from 53 per cent to less than 33 per cent (Figures 1.1 and 1.2). Paradoxically, although the pace of growth and structure of output changed dramatically after 1949, the composition of employment (and the population) remained predominantly agrarian. Although official data on the urban population is based on a changing definition and thus must be used with caution, China's development during the pre-reform period was nonetheless accompanied by an unusually slow pace of urbanization and structural change in employment (Chapter 2, Table 2.14).

Between 1953 and 1978 the labour force almost doubled to 401.5 million. However, agriculture absorbed almost two-thirds of that increase, and its share of employment only shrank from about 83 to 71 per cent of the total (Figures 1.3 and 1.4). The sharply declining trend in the product share of agriculture was accompanied by an unusually small reduction in its share in the total labour force. Consequently, in terms of production structure China resembled a relatively industrialized country, although its employment structure was similar to that of some of the world's least developed countries.

Regarding employment, a substantial proportion of the workforce was underemployed in the rural areas. During the Cultural Revolution, all farmers were strictly confined to crop production. They were not allowed to take up any jobs in the urban areas nor to diversify production activities in their own communities. Sideline production, high value-added crops and livestock production were restricted to a minimum. All these factors resulted in an enormous waste of labour resources and the stagnation of farmers' incomes.

The contradiction between the structure of output and the composition of the labour force raised several major issues for understanding Chinese economic development in the pre-reform

Figure 1.1   National income structure by sector, 1953

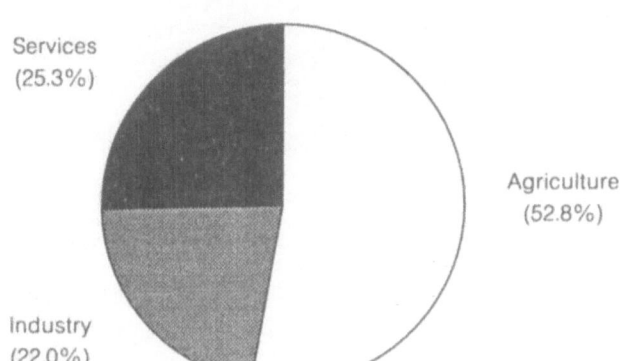

Source: Table 2.1.

Figure 1.2   National income structure by sector, 1978

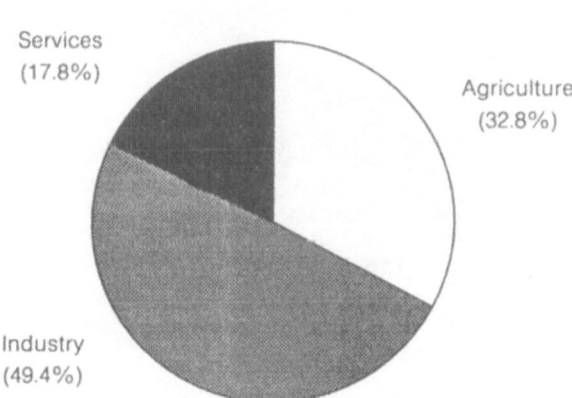

Source: Table 2.1.

Agricultural Policies and Economic Reforms

Figure 1.3   Employment structure by sector, 1953

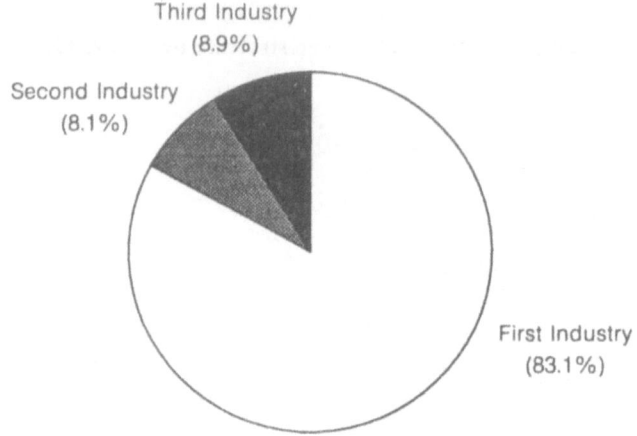

Notes: First industry mainly includes agriculture; second industry includes manufacturing and construction; third industry includes services and transportation.
Source: Table 2.14.

Figure 1.4   Employment structure by sector, 1978

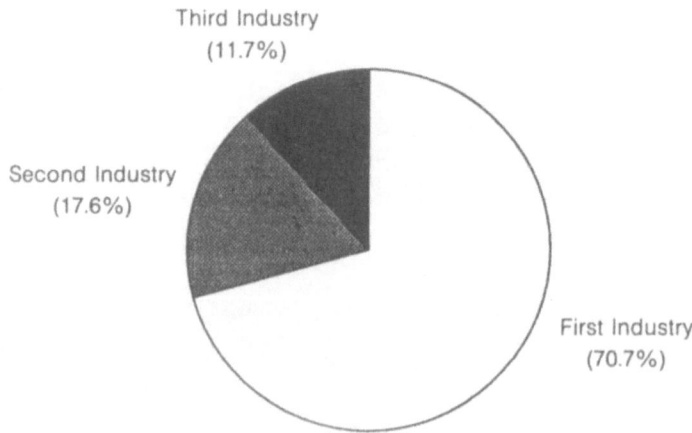

Notes: First industry mainly includes agriculture; second industry includes manufacturing and construction; third industry includes services and transportation.
Source: Table 2.14.

Table 1.1　Per capita peasant income, 1956–78 (yuan)[a]

Year	1956	1960	1965	1977	1978
Current prices	43.1	41.3	52.3	65.5	74.7
Constant prices	45.5	—	41.3	47.5	52.7[b]

Notes:
[a] The income only includes collective distribution.
[b] Estimated.
Sources: Agricultural Yearbook Compilation Commission (1981, p. 41); Ministry of Agricultural Policy Research Office (1980, p. 105); Ministry of Agriculture Commune Management Bureau (1981a, p. 117; 1981b, p. 13), SSB.

period. First, whereas industrial production and national income grew rapidly, the living standard of the peasantry stagnated, or even declined, between 1957 and 1978 (see Table 1.1).

Second, agricultural growth was highly unstable, particularly during the Great Leap Forward movement (1958–1962) and the Cultural Revolution (1966–1978). Whilst industrial output between 1953 and 1978 grew at an annual rate of 10.7 per cent, agriculture grew at only 2.7 per cent (Chapter 2, Table 2.10) and cereal output at only 2.4 per cent (Table 2.11; Yang Chien-pai and Li Hsueh-tseng 1980, p. 183). Since population growth averaged 2.0 per cent over this period, the margin for improvement in food consumption was rather modest for an economy in which per capita gross domestic product in constant prices tripled. Indeed, although China's industrial growth accelerated tremendously during the whole pre-reform period, it appeared that agricultural growth did little more than keep pace with the expansion of population.

One may argue that the slow pace of agricultural development in that period was explained primarily by China's land shortage. Whilst China's population grew by two-thirds between 1952 and the late 1970s, the area of arable land shrank by more than 10 per cent. Land reclamation programmes between 1957 and 1977 brought an additional 17 million hectares into cultivation, but this was more than offset by 27 million hectares of arable land used for building roads, housing, factories, and so on.

Arable land per agricultural worker also declined by half between the early 1950s and the late 1970s. Whilst the man:land ratio has been growing in virtually every developing country in Asia since the 1940s, increased pressure of population on the land was most intense in China

(Lardy, 1983). Under conditions of unchanging agricultural technology, an increased population on fixed or even decreasing land area can be expected, via the inexorable law of diminishing returns, to produce successively smaller increments of agricultural output. However, from the experience of the First-Five-Year Plan and the adjustment period between 1962 and 1965 after the disastrous Great Leap Forward, as well as the experience of the post-Mao period after 1978 (discussed in Chapter 2), Chinese agriculture could have developed much more rapidly if the policies had been appropriate.

1.3 INVESTMENT STRATEGY

As will be shown in detail later, the state allocated extremely modest investments either directly to agriculture or to the branches of industry that produce modern inputs for agricultural production. Moreover, the state's high indirect agricultural taxation (through low prices and mandatory procurement of grains and other crops) greatly curtailed reinvestment of internal resources within the farm sector. As a result, farm assets per agricultural worker remained extremely low. Thus the poor performance of agriculture may have been the consequence of an agricultural policy that tended to be myopic and extractive rather than developmental. Underinvestment in the mid-1960s when the pace of technical change in agriculture increased provides an example. Investment in agriculture was reduced just when its rate of return was rising due to new opportunities provided by new seed varieties, particularly rice. Skewed investment allocations increased the share of investment in industry (Chapter 3, Table 3.1), where the rate of return was lower than agriculture and declining. That distortion of investment allocation reduced the rate of growth of the economy below what could have been achieved.

Heavy industry was assigned almost 50 per cent of the total basic construction investment over the period 1953–78, whereas the share of investment in agriculture was only 12 per cent, and that in light industry was only 5.5 per cent (Table 3.1).

Whilst such a strategy might make sense over a short period (such as the First-Five-Year Plan) to establish a basic industrial infrastructure, pursued over the long term it implied a very crude view of the determinants of economic growth, paying too little attention to the dynamic effects of output on real income growth and to the efficiency of capital allocation.

Although the absolute numbers employed in the *secondary sector* (industry and construction) rose impressively, from 17.2 million in 1953 to 70.7 million in 1978 (Figures 1.3 and 1.4; and Chapter 2, Table 2.14), the structure of employment was not transformed. Heavy industry developed in its own way, ignoring the real market demand for its products. Consequently, huge amounts of products were permanently held in warehouses, whereas the market was short of agricultural products and other consumer goods produced by the light industrial sector, which depended largely on the farming sector for its inputs. Of course, allocative efficiency was far from optimal. Furthermore, a considerable amount of surplus agricultural labour (which was estimated to be more than 100 million farmers in 1978) could not be absorbed into the industrial sector which had already found it difficult to absorb the unemployed people in the cities and towns.

The outcome of such an industrialization strategy was as follows. Capital per worker grew rapidly over the long term, rising more than nine-fold from 1957 to 1978, while the number of industrial workers rose less than three-fold. The value of fixed assets per worker was vastly greater in industry than in agriculture. Chinese data report an annual average real growth of industrial output of about 9–10 per cent (compound) from 1957 to 1979, but at an increasing capital cost: in the state industrial sector gross value of industrial output per 100 yuan of fixed assets (original value) fell from 138 yuan in 1957 to 98 yuan in 1965 and still stood at only 105 yuan in 1975 (*Chinese Economic Yearbook*, 1982: Section 6.18).

To implement the heavy industry-oriented policy, the state had to supply the cities (industry) with enough food procured from the farming sector and set strict regulations on migration from the countryside to the urban areas. The resource transfer including food from the countryside took two major forms. The first was through direct agricultural taxation. This was relatively severe originally, but remained fixed in absolute terms so that its share in total farm income fell steadily. Agricultural tax fell from 13 per cent of net income, or 9.5 per cent of gross income, of the basic accounting units in 1958 to 4.3 per cent of net income, or 2.9 per cent of gross income in 1981 (SSB, 1981a, p. 124). The second was through indirect agricultural taxation in the form of low prices and mandatory state procurement of grain and other agricultural commodities. Since free marketing in rural areas was strictly regulated, official data on free market prices were not available for that period. However, as far as grain is concerned, free market prices used to be twice or three times as high as the state procurement prices paid to farmers. Even in 1980, when the procurement prices of grain had been greatly raised and free markets were deregulated, free market prices were still much higher than the procurement prices. According to a recent World Bank report, the net loss of farm revenue on mandatory grain sales was estimated to have been 23 billion yuan in 1988 (World Bank, 1991, p. 18), equivalent to 26 yuan per rural inhabitant and about 5 per cent of 1988 average rural per capita income. It has to be pointed out that resource transfer in the pre-reform period must have been far more severe in relative terms as the gap between state procurement prices and free market prices must have been much smaller in 1988 than in the pre-reform period.

In the pre-reform period, state procurement prices hardly covered the material costs of production. By delivering all of their surplus to the state, farmers were stripped of any possible cash income to improve their living standards and increase production capacity. The remaining part of the total output was usually not enough to feed the farmers. As a result, undernourishment or malnutrition was widespread through-out the countryside.

In principle, the compulsory procurement quota was set against the level of total production and the basic requirement for self-nourishment. In reality, the level of basic requirement was often treated as a residual after a fixed amount of quota. In addition, total state procurement was often raised by the so-called 'above-quota' procurement if harvests were good. However, if harvests were poor, state quotas were not usually adjusted downwards to allow more retention for self-nourishment. In the most revolutionary years, production teams were even 'educated' by the commune leaders to deliver much more than the quotas to the state through political praise; as a result, team members suffered much more severely. The revolutionary spirit was vividly demonstrated by the modern opera of 'Praise in the Dragon River' (*Long Jiang Song*) in the mid-1970s in which farmers were highly praised for delivering grains to the state.

The burden on farmers was not limited to compulsory delivery and low procurement prices. Excessive quality control and delayed payments for their products from the procurement stations incurred *invisible* costs. In Guangdong, for instance, the most difficult task for rice farmers was to meet the quality requirement of grain delivered to the procurement stations. In many areas, farmers had to carry their grain for a distance of 3 to 10 kilometres to the stations. If the grain could not meet the quality standards, which were often arbitrary and far too high, they had to carry the grain back and wait for later delivery. This tedious process of procurement incurred enormous labour costs which were fully borne by the production teams.

Because of the constraints of arable land as mentioned above, farmers had to raise yields by applying more and more current inputs such as fertilizers, insecticides and labour. As state procurement prices did not offer sufficient returns, farmers often found it impossible to buy these inputs due to the lack of cash income and rural credit. This created a vicious circle: low prices leading to low income, low income leading to low inputs, and low inputs leading to low production. Consequently, the income and living standards of the peasantry remained unchanged or increased only modestly for over two decades from 1956 to 1978 (Table 1.1).

Unfortunately, state extractions of grain and oils from the countryside at below-market prices did not lead directly to large state profits. The state commercial system incurred growing financial losses on the purchase and resale of grain and edible vegetable oils. The purchase of grain from the peasantry at low prices nonetheless contributed to a transfer from agriculture to other sectors of the economy, since maintaining low prices for basic food in urban areas was a central component of the state wage policy. After 1956 labour productivity in the state industrial sector rose at an average annual rate of almost 3 per cent, yet the state held real wages constant or declining for more than two decades. The combination of rising productivity and constant or even declining real wages allowed the profits of state-managed enterprises to soar from 11.4 to 49.3 billion yuan between 1959 and 1977 (SSB, 1960; SSB, 1981b). The fiscal system channelled most enterprise profits directly to the state budget. A constant or declining real wage strategy, however, was predicated partly on the continuous provision of basic foodstuffs to workers at low, fixed nominal prices. By 1961–62 this policy required state subsidies for the commercial sector. But over time, the real cost of the subsidy was held down by extracting grain from the countryside at low prices.

1.4 SELF-SUFFICIENCY AND COMPARATIVE ADVANTAGE

State intervention in agriculture was not limited to reducing its share of investment or extracting resources, especially grain and vegetable oils. State policy frequently inhibited the efficient use of the modest resources remaining within agriculture. The policy-makers tended to undervalue the law of value and the role of marketing. Periodic efforts were made to reduce specialization in production and marketing of agricultural products.

From 1949 to 1957 regions with resource endowments, which gave them a comparative advantage in animal husbandry or in the cultivation of economic crops, were encouraged to specialize. The state encouraged specialization not only through price and credit incentives, but also through the supply of staple foods to producers of economic crops. Prior to 1958 this supply was achieved through a combination of public and private marketing. Between 1953 and 1963 almost half of all grain collected by the state, through taxes and procurement, was resold in rural areas (Table 3.3).

After 1965, specialized production was abandoned in favour of a policy of local foodgrain self-sufficiency. Rural areas were still allowed to produce economic crops or raise animals, but only after they had achieved basic food self-sufficiency.

The policy of local self-sufficiency was not simply exhortative but it was also reinforced by other powerful policy instruments. First, the state reduced the share of foodgrain that it purchased or collected in taxes. In 1966, before the self-sufficiency policy was effectively underway, state procurement and taxes had surpassed the average of the First-Five-Year Plan and as a share of output were equal to the 25 per cent average of 1956 and 1957 (Chapter 3, Table 3.3). In the following decade procurement and taxes grew extremely slowly and in 1976 to 1978 averaged less than 48 million tons, up less than 10 per cent compared to 1966. Since grain production had increased by over a third, the share of output distributed directly by the state through the Ministry of Food declined from about 25 to 20 per cent. Moreover, the comparison of procurement prices and taxes in 1976–78 with those in 1956–57 understates the decline in marketing because private grain sales in the 1950s and presumably in the first half of the 1960s averaged several million tons per year, adding several percentage points to the share of foodgrain output that was 'marketed', whereas private grain sales from 1966 to 1978 were officially prohibited and probably quite small. This is reflected in the third column of Table 3.3, which shows marketed output declining from 30 per cent during the First-Five-Year Plan to 20 per cent in 1976–78 (see also the changes after 1978).

Not only was the share of grains procured by the state in 1977 and 1978 substantially smaller than that during the First-Five-Year Plan, the quantity of grain resold to the agricultural population grew by only a third between 1957 and the average of 1977 and 1978 while the agricultural labour force over the same period grew almost 70 per cent (Table 2.14). Since the total quantity of marketed grain reflects largely the degree of urbanization, the quantity of grain sold to peasants is a more accurate reflection of the extent to which marketing policy facilitated specialized production.

Inter-provincial transfers of grain, an important component of the policy of the 1950s facilitating specialization, were also curtailed substantially. By 1978 provincial exports had declined to less than 1.5 per cent of production compared to 4.7 per cent in 1953–56 and 2.8 per cent in 1965 (Lardy, 1983, Table 2.2). If grains, predominantly rice and soybeans, destined for international markets are excluded from inter-provincial transfers, the decline is even more dramatic: to 0.1 per cent in 1978 compared to 3.4 per cent in 1953–56 and 1.5 per cent in 1965.

Agricultural tax policy reinforced grain self-sufficiency. During the 1950s there had been a gradual tendency to allow the payment of agricultural tax in cash or crops other than grain, thus drawing peasants into specialized commercial crop production. Peasants in cotton-producing areas delivered cotton rather than grain to meet their tax obligation. In 1955, 8 per cent of the agricultural tax was paid in cotton (Lardy, 1983), a significant share for a crop that occupied only about 4 per cent of sown area. By the 1960s and well into the 1970s the state was more insistent on payment in grain. That posed a significant additional constraint on producing units since the agricultural tax in the mid-1970s amounted to about 13 million tons, almost one-third of all grain procurement by the state. Since grain could not usually be purchased on rural markets, it was necessary for all regions to grow grain to meet their own demands.

The pattern of growth after 1965 reflected the grain self-sufficiency policy and the declining profitability of commercial crop production. In the years 1966 to 1970 grain output increased at an average annual rate of 4.5 per cent, while non-grain crops and animals, except for tea, grew at lower rates: cotton 1.7 per cent, oilseed crops 0.6 per cent, sugar cane 0.1 per cent, sugar beet 1.2 per cent, live pigs 2.3 per cent, aquatic products 1.3 per cent, and tea 6.2 per cent (Yu Kuo-yao 1980, p. 29). The consequence of attempted self-sufficiency was declining allocative efficiency within agriculture and declining per capita consumption of many non-cereal foods.

Increased inefficiency was most evident in regions of northwest China which had a significant comparative advantage in the production of meat and other animal products. By the late 1960s these regions were forced to devote increased resources to grain production. Pasture lands were brought under grain production despite the lack of water resources adequate for growing field crops and the high probability of increased wind erosion. In many cases grain yields achieved on these lands were less than one-tenth of the national average and were obtained only at the expense of reduced production of high-value animal products.

Pursuit of increased grain production was not limited to predominantly pastoral regions but extended to areas previously specializing in non-grain crops, such as peanuts, oil-bearing seeds, cotton, and other economic crops. Consequently the area sown to these crops in regions with a significant comparative advantage frequently was reduced.

Increased inefficiency, however, was not confined to regions that were grain-deficient. While some regions were initially more than self-sufficient in grain, reduced opportunities to purchase non-grain crops from other regions led these grain-surplus regions to increase the share of their land allocated to producing cash crops. Because the yields of these crops were relatively low, that reallocation reduced efficiency.

1.5 INSTITUTIONAL CHANGES

Although he recognized agriculture as a major market for industrial goods and seemed to support, at least in official statement, a policy of concurrent growth, Mao Zedong sought to achieve agricultural growth primarily through organizational changes and to accelerate industrial development through a high level of state investment expenditures, financed largely through direct and indirect taxes on agriculture. He argued that the revolution in production relations could accelerate the expansion of production forces.

Organizational changes were considered necessary for building China as a modern advanced socialist country. The central issue of organizational change was to establish a system of public ownership. In the agricultural sector, the state tried to abandon private ownership of land and other major means of production. Following the land reform, peasants were immediately forced to set up mutual-aid organizations. In 1952, 40 per cent of the country's total peasant households belonged to mutual aid teams, and in 1954 they increased to 58 per cent. Simultaneously with swift growth of the mutual-aid teams, peasants were persuaded to organize 'semi-socialist' agricultural producers' co-operatives characterized by the pooling of land as shares and a single management. In 1952 there were only about 3600 agricultural producers' co-operatives, but they grew rapidly because they were considered to be more socialist. By the first half of 1955 their number increased to 670,000, embracing some 17 million households. In July 1955 Mao Zedong delivered his report 'The Question of Agricultural Cooperation'. Based on this report the Central Committee of the Chinese Communist Party adopted the 'Decision on the Question of Agricultural Cooperation' in October of the same year. Under the inspiration of these documents, a high tide of socialist co-operation on a magnificent scale appeared in the second half of 1955. By 1956 the organization of agricultural co-operation was in the main complete in China. By the end of 1956, 120 million peasant households (or 96 per cent of all the peasant households in China) had joined co-operatives. More than 100 million of them, or 88 per cent, joined the advanced agricultural producers' co-operatives (SSB, 1960).

The leadership expected that such radical institutional changes in the rural areas would pave the way for rapid growth in the rural economy. Presumably due to the socialist enthusiasm induced by land reforms and due to the emphasis on the roles of marketing, agricultural production increased rapidly during the First-Five-Year Plan (1953–57). The success in this period, therefore, encouraged Mao Zedong to accelerate the process of socialism through the establishment of the People's Communes.

Beginning in the summer of 1958, in the short space of a few months more than 740,000 agricultural co-operatives were merged and reorganized into over 26,000 large-scale People's Communes in which industry, agriculture, trade, education and military affairs were combined and government administration and commune management were merged. The communes embraced 120 million peasant households, or over 99 per cent of the total peasant households of all the nationalities in China (SSB, 1960).

After the establishment of the People's Communes in 1958, the peasants in the communes had to work under the supervision of the head of the commune. The state also practised forced procurement and distribution of some major crops such as grain, cotton and oil-bearing crops. Furthermore, the reward of individuals was not directly connected to their efforts. Everybody, regardless of his/her contribution to the commune's production, received the same reward. This system vastly discouraged the individual's incentive to work (for detailed discussion, see Chapter 3). Moreover, since the state's compulsory procurement quota of agricultural products, especially grain, cotton and pigs, was based on the previous production level or even on the plan target, the more the peasants produced, the more they had to deliver their products to the state. As a result, the communes refused to raise production because they could not benefit correspondingly. Consequently, although the state put much more emphasis on the production of grain as discussed above, the real performance was very disappointing. Foodgrain output declined dramatically after 1958. Accelerated by natural disasters, grain production in 1960 reached its lowest level since 1949. Between 1958 and 1960, the production of grain declined by 28 per cent. It was not until 1965 that total production recovered to the level of 1957. Per capita consumption declined even more. In 1960, per capita consumption decreased by 19.5 per cent compared to 1957. Because the urban population (non-agricultural population including those state employees working in the rural areas) was guaranteed adequate amounts of grain and oils by the state, per capita consumption in the countryside was much lower, especially for the poor areas in western China. Acute shortage of food in the late 1950s and early 1960s developed into a devastating famine. The extent of destruction caused by this disaster was reflected in both the low fertility and high death rate over the 'Three Difficult Years' of 1959–61. Yang Chien-pai has written that the population declined by more than 13 million between 1959 and 1961 (1980, p. 122).[1] One report shows that the mortality rate rose sharply in 1959 and reached a peak of 25.4 per cent in 1960 (Lardy, 1983), implying increased mortality in 1960 alone of over 9.5 million.

In 1962, some Party leaders (such as Liu Xiao-qu, Deng Xiaoping and Chen Yun) came to realize the serious situation in the rural areas and persuaded the government to lift the restrictions over the People's Communes. Consequently, the single commune accounting unit was transferred into the so-called 'system of three-level ownership (namely, the commune, the production brigade and the production team) with the production team as the basic accounting unit'.[2]

The system of household contract production was introduced. The private plot, which was abandoned in the Great Leap Forward, was re-introduced and enlarged.[3] Farmers were also allowed to trade their surplus products in the free markets after meeting the state procurement quota. Indeed, these methods were highly effective. Both the communes and the peasants were encouraged by the policies. Consequently, production soared again. By 1965 agriculture output recovered to the level in 1957. Unfortunately, Mao Zedong denied the achievements and the policies of this adjustment period. He still insisted on carrying on his programme of 'high-level socialism'. He launched the Great Cultural Revolution simply to punish the leaders who implemented the adjustment policies. His strategy, as revealed between 1966 to 1976, was to develop socialism through highly organized collectivization.

Beginning in 1966, the private plots were greatly reduced, and in some regions were totally abolished. The rural free markets were again considered 'capitalist' and closed. Those who traded their surplus products in the 'black markets' were severely punished, both economically and politically. Moreover, surplus labourers in the countryside were not allowed to take part in any activities in the urban areas. The production of sideline activities, cash crops and livestock was strictly controlled. Anybody who attempted to produce more than a certain amount of this production set by the commune (state) would be severely punished.[4] The state, represented by the commune, not only dictated peasants what to produce but when and how to produce. Since most of the commune leaders, who were state-salary employees, were incompetent in farming, their behaviour, usually corresponding to that of their superiors, was frequently irrational.

Political pressure and the irrational economic behaviour of the Party leaders greatly suppressed the incentives and the freedom of the Chinese peasantry. As will be discussed in detail in Chapter 3, the management structure of the people's commune unnecessarily increased the cost of production. It also made it impossible to monitor the performance of individual team members in the production process. Experience indicated that farmers worked hard throughout the year but about half of their income had to be deducted to cover the overheads of the commune, the production brigade and the production team. In a poor year, when the harvest of crops was low, team members might have no income at all after fulfilling tax and overhead obligations. The outcome of the policies was a modest pace of agricultural development and stagnating living standards of the peasantry for more than two decades.

1.6 CONCLUSIONS

After the civil war in 1949, the Chinese government practised a policy favouring heavy industry. A lion's share of state investment went to this sector. The agricultural and light industrial sectors received modest shares of investment compared to their roles in the national economy.

This unbalanced investment policy greatly reduced the allocative efficiency of resources and contributed to the slow pace of structural transformation of employment.

The state imposed a tremendous burden on the peasantry in terms of indirect taxation through mandatory procurement of grain and other agricultural products. Rigid regulations on migration restricted farmers' non-farm employment opportunities and increased population pressure on the land. Moreover, the state policy of grain self-sufficiency induced inefficiency of resource allocation within agriculture. Farmers could not produce according to regional comparative advantages because of forced production quotas and state intervention in food and agricultural marketing. As a result, the rural and urban disparity of income increased rather than decreased over time.

The political pressure for farmers to accept the system of high collectivization, the abolition of rural free markets, the irrational intervention by local bureaucracy and the excessive overhead cost in the commune were other factors to blame for slow agricultural growth and the stagnation of farmers' incomes.

2 Agricultural Reforms

2.1 INTRODUCTION

As discussed in Chapter 1, agricultural growth was modest and the living standards of farmers stagnated for more than two decades before 1978 although national income grew rapidly and the production structure was greatly transformed. By contrast, economic reforms in the 1980s have resulted in rapid agricultural growth and vast improvements in the living standards of the whole rural population: per capita income and food production have increased steadily; the share of labour engaged in the traditional agricultural sector, especially crop production, has been substantially reduced; the production of non-food crops has experienced much higher growth; and rural industries have become the most important production and income-generating sector of the rural economy in most fast-growing regions, especially the coastal areas.

Detailed discussion of reform policies will be presented in Chapter 3. This chapter describes the dramatic changes brought about by these policies in terms of production, incomes and structures of the national economy and employment. Section 2.2 presents the statistics of income growth and the improvements of living standards for the population, especially for the rural population. It also presents a set of indicators on production growth and the structural changes in production and employment.

To understand the extent of changes in production and living standards in rural China during the reforms in a more visible way, a personal experience of the changes in a remote village is presented in section 2.3. The village is considered to be an *average* village in eastern Guangdong. Thus the changes in that village are considered to be representative for that part of the province. In comparison with other parts of Guangdong, the eastern part is below average. The most developed areas in the province are around the capital city, Guang-zhou, where living standards are much higher than in the eastern areas. However, Guangdong is one of the most prosperous regions in China and has benefited most from the reforming policies during the last decade. Thus Guangdong is not representative of the whole country although significant and similar changes have undoubtedly occurred throughout the country since the inception of reforms, as indicated by the statistics presented in Section 2.2.

2.2   GROWTH AND STRUCTURAL CHANGES AFTER 1978

Many economic indicators suggest that economic growth has been much faster since the reforms than before the reforms. Significant changes brought about by economic reforms are reflected in two sets of indicators. One includes income growth and the improvements in living standards, especially for the rural population. The other includes production growth and the structural changes in production and employment.

2.2.1   Income growth

Since 1978, agricultural production and the whole national economy have developed rapidly. As shown in Table 2.1, the national income index measured at constant price increased by almost 9 per cent per annum between 1978 and 1989, or more than 3 per cent higher than the average growth rate between 1953 and 1978 (Table 2.1 and Figure 2.1). At current prices, national income increased from 301 billion yuan in 1978 to 1312 billion yuan by 1989. The following three years, 1990–2, saw even higher growth rates.

Rapid economic growth has been accompanied by significant improvements in the living standards of people, especially the rural population. Table 2.2 reveals these changes. First of all, per capita disposable incomes in both the urban and rural areas increased rapidly. Secondly, for the first time since independence, farmers' incomes increased faster than workers' incomes, reducing the urban–rural gap which was wide and had been widening in the pre-reform period. Lardy (1983), and Yao and Colman (1990) concluded that there was little real income growth for the Chinese peasantry for more than two decades between 1957 and 1978. The gap between the urban and rural incomes was widening throughout the pre-reform period, not because of the increase in real urban wages but because of more employment opportunities which were only available to the urban citizens. Increased per capita urban income was mainly achieved by a decreasing dependent ratio.

Table 2.1   National income by sector (100 million yuan)

Year	National Income	Nat. Inc Index.	National Income	As % of national income		
	Current Prices	Constant Prices	Index Per Cap.	Agri. (%)	Industry (%)	Others (%)
1952	589	100.0	100.0	57.7	19.5	22.8
1953	709	114.0	111.5	52.8	22.0	25.3
1960	1220	199.1	172.9	27.2	46.3	26.5
1965	1387	197.4	156.4	46.2	36.4	17.4
1970	1926	294.6	204.0	40.4	41.0	18.6
1971	2077	315.3	212.6	38.9	42.9	18.2
1972	2136	324.3	213.8	37.8	44.1	18.1
1973	2318	351.2	226.3	38.2	44.0	17.8
1974	2348	355.2	224.7	39.3	43.2	17.5
1975	2503	384.7	239.3	37.8	46.0	16.2
1976	2427	374.5	229.7	38.7	45.6	15.7
1977	2644	403.7	244.3	34.5	47.8	17.7
1978	3010	453.4	270.7	32.8	49.4	17.8
1979	3350	485.1	285.9	36.6	48.6	14.8
1980	3688	516.3	300.7	36.0	48.9	15.1
1981	3941	541.5	311.0	38.3	46.7	15.0
1982	4258	585.8	331.6	40.5	45.8	13.8
1983	4736	644.2	361.3	40.6	45.1	14.3
1984	5652	731.9	406.6	39.8	44.5	15.7
1985	7020	830.6	456.7	35.5	45.1	19.4
1986	7859	894.5	486.3	34.6	45.5	19.9
1987	9313	985.7	529.9	33.9	45.8	20.4
1988	11738	1097.2	580.4	32.5	46.1	21.3
1989	13176	1137.2	592.3	31.9	47.4	20.7
1990	14384	1195.5	601.0	34.8	46.0	19.3
1991	16117	1286.4	646.7	32.7	47.8	19.5
Annual growth rates						
Pre-reform period						
1965/53	5.75	4.68	2.86	−1.10	4.29	−3.07
1978/65	6.14	6.61	4.31	−2.61	2.37	0.21
1978/53	5.95	5.68	3.61	−1.89	3.29	−1.38
Reform period						
1984/78	11.07	8.31	7.01	3.31	−1.72	−2.15
1988/84	20.05	10.65	9.31	−4.94	0.90	8.03
1991/78	13.78	8.35	6.93	−0.02	−0.25	0.69

Notes:
(1) National income is measured at current prices; indices are at constant prices.
(2) 'Others' includes all the other sectors of the national economy.
Sources: *CSYB* (Chinese edition) (1992), pp. 34–5 and Table 2.14.

Figure 2.1   Per capita national income index, 1952–91 (1952 = 100)

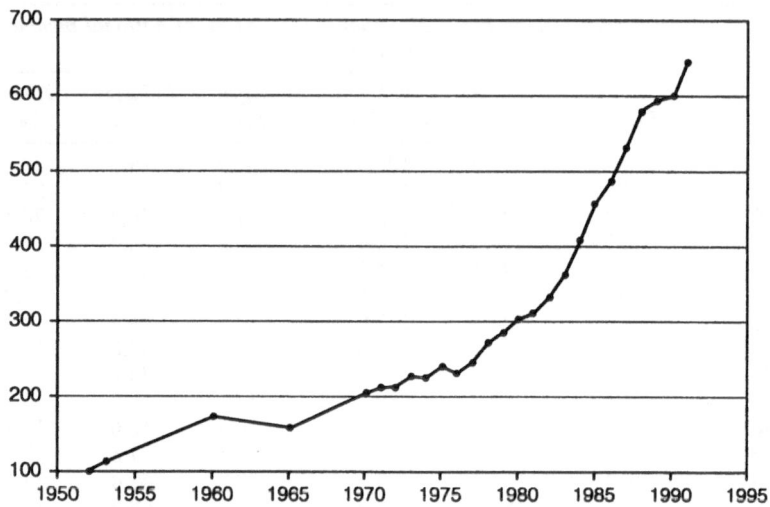

Source: Table 2.1.

In the reform period, urban incomes have been increasing faster than before. Between 1978 and 1989, the real per capita income of the urban population increased by 6 per cent per annum (by 13.4 per cent at current prices). Per capita housing space increased also by almost 6 per cent per annum (Table 2.2). In real terms, urban per capita expenditure increased by more than 5 per cent per annum. However, these high growth rates were outstripped by the same indicators for the rural population. Per capita income for the rural population increased by more than 8 per cent over the same period (14.7 per cent at current prices). Rural per capita expenditure and housing space all increased by more than 7 per cent per annum. By 1989, per capita housing space in the rural area was more than $17\,m^2$, compared to only $6.6\,m^2$ in the urban area.

A more direct indicator of improved living standards is the ownership of consumer durable goods. According to the SSB annual household surveys, the ownership of durable goods, such as TV sets, watches, bicycles, washing machines and radios, reflects how an average Chinese household moves away from subsistence to modernity. Table 2.3 presents the average numbers of different durable consumer goods per 100 people. It indicates that the level of

Table 2.2 Per capita income and housing space of worker and peasant families (yuan and m²)

	At current prices								Annual growth rate (%)		
	1978	1984	1985	1987	1988	1989	1990	1991	1978/84	1985/91	1978/91
Urban family											
Income (yuan)	316.0	607.6	685.3	916.0	1119.4	1260.7	1387.3	1544.3	7.7	7.0	7.8
Housing (m²)	3.6	4.9	5.2	6.1	6.3	6.6	6.7	6.9	5.3	4.8	5.1
Expenditure	383.0	619.3	673.2	884.4	1104.0	1211.0	1278.9	1453.8	4.5	6.1	5.6
Rural family											
Income (yuan)	133.6	355.3	397.6	462.6	544.9	601.5	686.3	708.6	13.4	1.3	7.5
Housing (m²)	8.1	13.6	14.7	16.0	16.6	17.2	17.8	18.5	9.0	3.9	6.6
Expenditure	132.0	278.7	317.4	398.3	476.7	535.4	584.6	619.8	8.9	3.0	6.5

Notes:
(1) The above figures are obtained from sample surveys conducted by the SSB.
(2) Per capita incomes for rural families are net incomes; per capita incomes for the urban families are those of living expenses only.
(3) Growth rates are calculated at constant prices. The expenditure growth rates are derived from the *CSYB* (1992), p. 277 which gives real expenditure index. The differences of the growth rates of expenditures at current and constant prices are used to derive the real growth rates of incomes.

Sources: 1978 and 1984 *BR* (1985c), p. 17; *CSYB* (1989), pp. 289–324; (1992), p. 282, p. 306 and p. 277.

consumption for these goods has increased enormously. Some durable goods such as TV sets, cameras, washing machines, refrigerators and video recorders tended to be solely consumed by the urban population a few years ago. As farmers' incomes have increased and the construction of new houses has reached a very high level, rural people have begun to buy large numbers of TV sets, *famous brand* bicycles, and cassette recorders. The market for *normal* bicycles, watches and radios seems to be saturated in many parts of rural China.

Table 2.3   Numbers of durable goods per 100 people

Durable	1978	1980	1985	1987	1988	1989	1990	1991
Sewing machines	3.5	4.7	9.4	11.0	11.8	12.2	12.3	12.6
Watches	8.5	12.9	34.5	42.8	47.0	50.1	—	—
Bicycles	7.7	9.7	21.4	27.1	30.4	32.8	34.2	36.2
Electric fans	1.0	1.4	6.1	10.4	13.4	15.6	17.6	19.8
Washing machines	—	—	2.9	5.3	6.8	7.8	8.4	9.2
Refrigerators	—	—	0.4	1.1	1.8	2.3	2.6	3.0
TV sets	0.3	0.9	6.7	10.7	13.2	14.9	16.2	17.8
Cassette recorders	0.2	0.5	3.5	6.5	8.3	9.6	10.4	11.3
Radios	7.8	12.1	23.1	24.1	23.9	23.6	22.0	20.2
Camera	0.5	0.6	1.1	1.5	1.7	1.9	2.0	2.1

Notes: Data are obtained from the national household survey for both the rural and urban families.
Sources: *CSYB* (Chinese edition) (1990), p. 294; (1992), p. 280.

The data in Table 2.3 are not differentiated for the rural and urban households. However, according to similar data compiled by Nolan, the gap of ownership of consumer durable goods between the rural and urban people has been significantly reduced. For example, between 1978 and 1988, the ratios of the numbers of consumer durable goods owned per 100 urban people over the numbers of such goods owned per 100 rural people were reduced from 3.58 to 2.74 for sewing machines, from 5.41 to 2.45 for bicycles, from 7.33 to 2.1 for watches, from 3.96 to 1.65 for radios, and from 13.0 to 5.86 for TV sets (Nolan, 1991, Table 3).

Increased incomes have been accompanied by significant changes in the composition of household expenditure. The change in the expenditure pattern for an average Chinese rural household has been much more profound than for an average urban household. This is particularly reflected in the share of total expenditure allocated to food consumption. International experiences suggest that a reduced share of food expenditure in total expenditure is an indirect indication of improved living standards. In China, the share of food expenditure in total expenditure for an urban household used to be much lower than that for a rural household. In 1978, for instance, farmers had to spend more than 67 per cent (Table 2.5) of total expenditure on food, while urban dwellers spent only 57 per cent (Table 2.4). Although the share of food expenditure of the urban households decreased by 3 per cent from 1978 to 1989, it decreased by more than 13 per cent for the rural households. Thus, by 1989, there was no difference in food expenditure shares between the urban and rural households. It is necessary to stress that urban people have tended to spend more of their incomes on high-quality foods because they do not need to spend as much on housing as their rural counterparts, but the reduced gap between the shares of food expenditure of the urban and rural households is a strong indication that farmers have benefited relatively more during the reforms. A more important implication is probably that most Chinese farmers are not on subsistence any more after ten years of economic reforms, although there is still evidence that farmers in some remote and underdeveloped areas are still very poor.

Vast improvements in living standards in rural China may disguise a worrying inequality in income distribution between the rich and poor. Bhalla has done a thorough analysis of this issue (Bhalla, 1992). He compared inequality of development in terms of income distribution, access to education, access to health services and access to technology. According to his results, income inequality for the urban population was reduced between 1978 and 1983 but slightly increased for the rural population. The GINI coefficient calculated decreased from 0.16 in 1978 to 0.14 in 1983 for the urban sector but increased from 0.22 to 0.26 for the rural sector over the same period (Bhalla, 1992, Table 6.1).

The recent study conducted by the Research and Development Centre of the State Council, the official think-tank of the government, confirms Bhalla's conclusion. The GINI coefficient of farmers' income increased steadily from 0.21 in 1978 to 0.30 in 1988 (Table 2.6). Although similar studies are not available, there is clear evidence that inequality in the rural sector has been increasing since 1988. The widening gap appears to be explained by the uneven development of rural industries which have become the most important source of farm incomes.

Table 2.4 Makeup of annual expenses for an average worker family

	Expenses at current prices				Percentage			
	1978	1984	1989	1991	1978	1984	1989	1991
Total expenses	311.2	559.4	1210.9	1453.8	100.00	100.00	100.00	100.00
I Commodity expenses of these:	279.6	514.3	1099.9	1294.9	89.86	91.93	90.83	89.07
1 food	178.9	324.2	660.0	782.5	57.50	57.97	54.50	53.82
2 clothing	42.2	86.9	149.2	199.6	13.58	15.53	12.32	13.73
3 fuel	8.4	9.2	18.2	25.1	2.70	1.66	1.50	1.72
4 medicine	3.1	3.4	16.0	23.0	1.00	0.60	1.32	1.58
5 daily necessities	26.2	50.6	134.0	139.8	8.41	9.06	11.07	9.62
6 recreation items such as books and newspapers	16.1	30.5	80.6	68.7	5.17	5.44	6.66	4.72
II Other expenses of these:	31.6	45.1	111.1	159.0	10.14	8.07	9.17	10.93
1 rent	6.0	7.8	8.8	10.7	1.93	1.38	0.73	0.73
2 electricity and water	4.2	6.2	16.2	24.2	1.35	1.11	1.34	1.66
3 transport and postal services	5.2	8.3	11.4	19.7	1.67	1.47	0.94	1.36
4 tuition and cultural activities	4.7	6.6	28.8	40.0	1.50	1.19	2.38	2.75

Sources: 1978, 1984 BR (1985a); 1989 and 1991, CSYB (Chinese edition) (1990), p. 300; 1992, p. 284.

Table 2.5 Makeup of annual expenses for an average peasant family (yuan and percentage)

	Expenses at current prices				Percentage			
	1978	1984	1989	1991	1978	1984	1989	1991
Total	116.1	273.8	535.4	619.8	100.00	100.00	100.00	100.00
I On commodities	112.9	267.3	500.1	571.2	97.30	97.60	93.41	92.16
1 food	78.6	161.5	289.6	352.3	67.70	59.00	54.09	56.84
2 clothing	14.7	28.3	44.4	51.0	12.70	10.40	8.29	8.23
3 fuel	8.3	15.0	23.5	26.8	7.10	5.50	4.39	4.33
4 housing	3.7	32.1	77.1	68.9	3.20	11.70	14.40	11.12
5 daily necessities, medicine and others	7.6	30.3	65.6	72.2	6.60	11.00	12.25	11.65
II On culture and services	3.2	6.5	35.3	48.6	2.70	2.40	6.59	7.84

Sources: 1978, 1984, BR (1985c), p. 18; 1989 and 1991, CSYB (Chinese edition) (1990), p. 316; 1992, p. 310.

33

34   *Agricultural Policies and Economic Reform*

Table 2.6   The GINI coefficients of farmers' incomes, 1978–88

Year	1978	1980	1983	1985	1986	1987	1988
Coeff	0.2124	0.2366	0.2459	0.2635	0.2848	0.2916	0.3014

Sources: The Research and Development Centre of the State Council (1989), p. 167; Guo Shutian *et al.* (1992), p. 87, Table 6–5.

Table 2.7   Regional income disparities in rural areas (yuan/p.a.)

Regions	Eastern		Central		Western	
	1986	*1989*	*1986*	*1989*	*1986*	*1989*
Income per capita	496	766	399	529	328	461
Engel Coefficient (%)	52	51	57	55	63	59

Sources: *China Rural Statistical Yearbook*, 1987 and 1990; Guo Shutian *et al.* (1991), p. 267, Table 8–6.

As rural industries have a very clear geographical development pattern, so does income distribution. Table 2.7 indicates the degree of income inequality between different parts of China. In 1986, rural per capita income in the eastern part of the country was about 25 per cent higher than in the central region, and 50 per cent higher than in the western areas. By 1989, the differentials had increased to over 40 per cent and over 66 per cent respectively. In 1989, rural income per capita in Shanghai (the richest region) was 1208 yuan, whereas in Gansu (the poorest region) it was only 290 yuan. The former was four times as much as the latter.

Table 2.8   Urban–rural income disparities (yuan per capita)

Year	*1978*	*1980*	*1983*	*1984*	*1985*	*1988*	*1989*	*1990*
Rural	134	191	310	355	398	545	602	686
Urban	316	439	526	608	685	1119	1261	1387
Urban/Rural Ratio	2.36	2.29	1.69	1.71	1.72	2.05	2.09	2.02

Sources: *China Statistics Abstract* (1990); *China Statistics Yearbook* (1990); Guo Shutian *et al.* (1991), p. 97, Table 6–13.

The income distribution of the urban and rural people has had a mixed development. Urban–rural income disparity was significantly reduced during the earlier years of economic reforms. The urban:rural per capita income ratio declined from 2.36 in 1978 to 1.69 in 1983. Since 1983, the urban–rural gap has been gradually widening again. By 1990, the urban:rural income ratio had increased to 2.02 (Table 2.8). The reversed trend of urban–rural income disparity reflects the lack of a genuine commitment from the government to agricultural development and farmers' incomes. It reveals that even during the reforms, the state is still very much biased towards the urban sector. This leads many scholars to believe that the prospects for Chinese agriculture are largely dependent on the sincere willingness of the government to support agriculture and farmers (see Sicular, 1993; Yao and Colman, 1990).

There is no doubt that income inequality in rural China has been increasing since the reforms but, in general, the degree of inequality is still relatively low compared to other developing countries. For example, the GINI coefficient in rural India is generally much higher than in rural China. According to Bhalla's calculations, the GINI coefficient of rural India was 0.33–0.39 during the 1970s although income inequality had been reduced from the late 1960s (ibid).

Increased income inequality in rural China may be inevitable during economic reforms for a number of reasons. Firstly, China is such a large country that there are enormous geographical differences across regions. Rapid development in regions well-endowed with resources, both natural and human, will take considerable time to trigger development in other parts of the country, which are remote and suffer from acute shortage of transportation, communications, capital and human resources. Secondly, income distribution, especially intra-regional income distribution before economic reforms, was unusually equal. Economic reforms which have been characterized by decollectivization, market liberalization and the development of rural enterprises will inevitably increase disparity. Areas which are close to cities and towns have easier access to the urban markets. Those which have richer endowments of natural and human resources are more efficient in production. This has been a characteristic of development in the rural areas along the eastern coast and around the large metropolitan centres in the inland areas. In these areas, agricultural productivity is much higher and the development of non-farm enterprises much faster.

A critical question is: to what extent can income inequality be tolerated? As indicated above, inequality in China in the early 1980s was not particularly skewed. More importantly, faster growth of income in the richer areas had not been achieved by squeezing the incomes of the poor areas. Recent statistics indicate that most farmers in China have moved from lower income to higher income brackets. In 1978, more than 97 per cent of households had a per capita income of less than 300 yuan. By 1989, only 15 per cent of households had a per capita income of less than 300 yuan but more than half of the households had per capita income in excess of 500 yuan (Table 2.9). Overall, it can be concluded that economic reforms in rural China have made all farmers richer, but some richer than others.

In terms of access to social services, especially health care, there has been an increasing concern that the household responsibility system in China may have reduced their effectiveness due to the abolition of the communes. Before economic reforms, China was very successful in the areas of basic education and health as was reflected in the dramatic improvement in life expectancy and popularization of primary education.

Comparing the experiences of health care between India and China, Bhalla (1992) suggests that the success of the Chinese in providing adequate health care at the grass-roots level during the Mao period was due, in part, to the generally egalitarian economic and social policies, and in part to the group motivation and commitment nurtured in the collective organization in the rural areas. In the post-Mao period, the private health services have supplemented the public ones. Although the former involve somewhat higher costs of medical care (e.g., doctors' fees, hospital charges and costs of drugs) they are of better quality than the public services. A revealed preference of peasants for private health care suggests that the better-off rural people can afford these services. However, the post-Mao period has witnessed growing inter-household income inequalities especially in the rural areas. Therefore, in future the poorer households are unlikely to enjoy as much access to health services as the richer ones.

Nolan (1991) has a similar analysis on the impact of reforms on health services in China. He also concludes that income inequality has risen in the post-Mao period, and the number of part-time rural doctors ('barefoot doctors') declined, but the critical indicators of health services, such as infant mortality and death rate, are not positively correlated with incomes. For instance, average infant mortality decreased by 0.6 per 1000 from 6.9 per 1000 in 1978 to 6.3 per 1000 in 1989 for the five poorest provinces, but it decreased by only 0.1 per 1000 from 5.9 to 5.8 per 1000 over the same period for the five

Table 2.9 Grouping of peasant households by income

Per capita net income (yuan)	Share of household number of each income group as % of total number of households								
	1978	1980	1985	1986	1987	1988	1989	1990	1991
< 100	33.3	9.8	1.0	1.1	0.9	0.5	0.6	0.22	0.40
100–150	31.7	24.7	3.4	3.2	2.4	1.5	1.3	0.51	0.77
150–200	17.6	27.1	7.9	7.0	5.0	3.3	2.8	1.29	1.56
200–300	15.0	25.3	25.6	21.8	17.5	13.5	10.9	6.57	6.64
300–400	> 300 ⌉	8.6	24.0	21.7	21.3	13.5	15.6	11.99	11.08
400–500	2.4	2.9	15.8	16.5	17.2	16.7	15.6	14.37	13.35
500–600	> 500 ⌉ 1.6 ⌋	> 500 ⌉ 1.6	9.1	> 500 ⌉ 28.7	12.00	13.30	13.40	13.99	13.00
600–800			8.02		12.87	16.45	17.63	20.83	20.90
800–1000			2.93		5.47	7.81	9.54	12.45	12.83
1000–1500			1.89		4.10	6.69	8.86	12.20	12.99
1500–2000			0.26		0.88	1.79	2.55	3.47	3.89
>2000			0.16		0.38	0.93	1.46	2.11	2.59

Notes: Net incomes are measured at current prices.
Sources: CSYB (Chinese edition) (1990), p. 312; (1992), p. 306.

richest provinces (which have per capita income more than twice as high as the poorest provinces: see Nolan, 1991, Table 17).

There are several factors which may help to explain this controversial finding. Firstly, reforms have not only increased the incomes of the rich, but also those of the poor. The ability of the poor to afford better services must have been increased. Secondly, although the number of barefoot doctors declined during the reforms, they have been replaced by fewer but better qualified doctors in the local hospitals and clinics. Thirdly, even in the Mao period, medical care in the rural areas was not totally free and the quality and availability of services largely depended on the public funds available, which were very restricted in most areas. By contrast, health care and other public services, such as education and care for the elderly, have been improved in many rural areas since the reforms, thanks to rising incomes and public funds. During the reform period, numerous new schools and homes for the elderly have been established and the working conditions of primary school teachers (most of whom are collective employees) have been greatly improved in all the fast-growing areas, owing to increasing investment by the townships and villages. The investment funds come from various sources: profits of rural enterprises, donations from overseas Chinese, and most importantly, the contribution of individual farmers (either through compulsory local taxes or voluntary donations).

2.2.2 Structural changes

The success of economic reforms is not only revealed by rapid income growth, but also by rapid production growth and significant structural changes in production and employment.

Agricultural production experienced an unprecedented upsurge shortly after the reforms. During 1978 to 1984, total agricultural output increased by 7.62 per cent (Table 2.10) per annum, compared to only 2.68 per cent between 1953 and 1978. As a result, the share of agricultural output in total national income for the first time after independence had a positive growth rate of more than 3.31 per cent between 1978 and 1984. During 1953 to 1978, the share of agricultural output in the national economy decreased by almost 2 per cent per annum.

Due to structural changes and a number of other factors (such as low investment and low profitability), the growth rate of agricultural output slowed down after 1985, when the outputs of grain and

Table 2.10 Growth rates of agriculture and industry at 1952 prices (average annual rate, %)

	Pre-reform period				Reform period			
	1953–7	1958–65	1966–78	1953–78	1978–84	1984–9	1978–89	1989–91
Total agricultural and industrial values	10.05	2.75	7.55	7.97	8.70	13.14	10.91	9.80
Agriculture	4.89	1.00	2.47	2.68	7.62	3.92	5.92	5.65
Industry	15.09	3.58	9.68	10.71	9.55	15.89	12.39	11.08
Light Industry	9.67	4.99	8.10	8.63	12.32	16.80	14.33	11.88
Heavy industry	22.74	2.28	10.93	12.96	7.23	15.00	10.70	10.24

Source: CSYB (Chinese edition) (1990), p. 57; (1992), p. 55.

cotton, the two most important crops in China, plummeted by 6 per cent and 50 per cent respectively from their record levels in 1984. Agricultural output between 1984 and 1989 increased by less than 4 per cent per annum (Table 2.10). Although much lower than that between 1978 and 1984, it was still higher than that in the pre-reform period. On average, agricultural growth rate was almost 6 per cent per annum for the whole reforming decade of 1978–89, or 3.24 per cent higher than the average growth rate in the previous two decades.

Rapid agricultural growth in the earlier stage of reform (1978–84) is also revealed by the dramatic growth of grain crops, cash crops, meat and fish. Between 1978 and 1984, total grain output increased by more than one-third, from 305 million to 407 million tons, registering an annual growth rate of almost 5 per cent (Table 2.11). Over the same period, cotton production grew by more than 18 per cent per annum, oil-seeds by almost 15 per cent, sugar crops by 12 per cent, meat by 10 per cent and fish by almost 5 per cent. Although crop production, especially grain and cotton, suffered a severe setback between 1985 and 1988, the growth rates of these major agricultural goods were still much higher in the reform period than in the pre-reform period. Between 1978 and 1989, grain production had an average annual growth rate of 2.83 per cent, or 0.4 per cent higher than that between 1953 and 1978. In terms of per capita production growth, it was even higher in the reform period, with an annual rate of 1.56 per cent, as compared to only 0.44 per cent in the pre-reform era. This was partly due to a lower population growth rate in the later period (1.25 per cent as opposed to 1.99 per cent: Table 2.11).

The average annual growth rates of cash crops, meat and fish were much higher than that of grain in the whole reform period, with cotton at 5.2 per cent, oil-seeds 8.6, sugar crops 8.4, meat 10.7 and fish 8.6 per cent between 1978 and 1989. These records contrast sharply to those in the pre-reform era. Between 1952 and 1978, the average annual growth rates of cotton, oil-seeds, sugar crops, meat and fish were only 0.25, 2.8, 3.4, 3.5 and 3.4 per cent respectively (Table 2.12). It is ironic that grain production was overemphasized in the past but grain output did not grow rapidly, whilst at the same time the production of non-grain commodities was suppressed. In the reform period, non-grain agricultural goods were allowed to expand, but at the same time, grain production was growing faster than before. As will be discussed in Chapter 3, this paradox is largely explained by increased allocative efficiency and productivity, rather than by greater physical inputs. In fact, the input of land had been declining over the whole reform period.

Table 2.11 Grain–population balance

Year	Grain Output (mil. ton)	Population (million)	Grain output per capita (kg)	Net grain imports (1000 tons)
1949	113.20	541.60	209.01	0.00
1950	132.10	551.90	239.35	−1158.90
1951	143.70	563.00	255.24	−1971.10
1952	163.90	574.80	285.14	−1528.70
1953	166.80	587.90	283.72	−1811.60
1954	169.50	602.66	281.25	−1680.00
1955	183.90	614.65	299.19	−2051.20
1956	192.70	628.28	306.71	−2502.00
1957	195.10	646.53	301.76	−1925.80
1958	200.00	659.94	303.06	−2659.90
1959	170.00	672.07	252.95	−4155.50
1960	143.50	662.07	216.74	−2654.10
1961	147.50	658.59	223.96	4454.70
1962	160.00	672.95	237.76	3892.10
1963	170.00	691.72	245.76	4461.90
1964	187.50	704.99	265.96	4749.30
1965	194.50	725.38	268.14	3988.70
1966	214.00	745.42	287.09	3552.80
1967	217.80	763.68	285.20	1707.50
1968	209.10	785.34	266.25	1995.10
1969	210.90	806.71	261.43	1548.80
1970	239.90	829.92	289.06	3240.50
1971	250.10	852.29	293.44	555.70
1972	240.50	871.77	275.88	1830.60
1973	264.90	892.11	296.94	4234.80
1974	275.20	908.59	302.89	4477.40
1975	284.50	924.20	307.83	928.90
1976	286.30	937.17	305.49	601.80
1977	282.70	949.74	297.66	5687.80
1978	304.70	962.59	316.54	6955.30
1979	332.10	975.42	340.47	10704.50
1980	320.50	987.05	324.70	11811.00
1981	325.00	1000.72	324.77	13551.40
1982	354.50	1016.54	348.73	14865.70
1983	387.30	1030.08	375.99	11472.00
1984	407.30	1043.57	390.29	7210.30
1985	379.10	1058.51	358.14	−3320.00
1986	391.50	1075.07	364.16	131.00
1987	404.70	1093.00	370.27	13000.00
1988	399.30	1110.26	359.65	11400.00
1989	414.40	1127.04	367.69	10350.00
1990	446.24	1143.33	390.30	7890.00
1991	435.29	1158.23	375.82	2590.00

(cont. overleaf)

Table 2.11 *continued*

Year	Grain Output	Population	Grain output per capita	Net grain imports
Annual growth rate (%):				
Pre-reform period				
1949–1952	13.13	2.00	10.91	—
1952–1957	3.55	2.38	1.14	—
1957–1962	−3.89	0.80	−4.66	—
1953–1965	1.29	1.77	−0.49	—
1965–1978	3.51	2.20	1.28	—
1953–1978	2.44	1.99	0.44	—
Reform period				
1978–1984	4.96	1.21	3.70	—
1984–1988	−0.49	1.56	−2.02	—
1978–1991	2.78	1.43	1.33	—

Notes: ' + ' Net Import; '−' Net Export. The USDA's data are not consistent with the Chinese official data.
Sources: *ZGNCJJTJDQ* (1949–1986), 1989; *ZGNYTJZL*, various issues 1987–91; *CSYB* (1992), p. 77 (population); p. 357 (grain output); p. 635 and p. 638 (export and import, 1990–91). Net import data for 1988–9 are from the US Department of Agriculture, Foreign Agriculture Service, *World Grain Situation and Outlook*, various issues.

Since 1978, rapid production growth has been accompanied by significant changes in the structures of production and employment. In the pre-reform period, economic growth was achieved by developing heavy industry at the expense of agriculture and the light industry. As a result, economic growth was not accompanied by significant changes in the structure of employment and production. By 1978, the share of heavy industrial output was as high as 42.8 per cent of the total agricultural and industrial output while the shares of agricultural and light industrial output were only 24.8 per cent and 31.2 per cent respectively (Table 2.13). In the same year the share of heavy industrial output in total industrial output was 57 per cent whilst that of light industrial output was only 43.1 per cent. By 1989, the structure of the three production sectors became more balanced. The share of heavy industrial output in total agricultural and industrial output decreased by 3.4 per cent from 1978 to 1989, but the share of light industrial output increased by 5.3 per cent. The share of agricultural output in total agricultural and industrial output increased for the first time after

Table 2.12   Outputs of major agricultural products

Year	Cotton (mil. ton)	Oil-bearing Crops (mil. ton)	Sugar-bearing Crops (mil. ton)	Meat (mil. ton)	Fish (mil. ton)
1953	1.17	3.86	7.71	3.39	1.90
1965	2.10	3.63	15.38	5.51	2.98
1970	2.28	3.77	15.52	5.97	3.19
1978	2.17	5.22	23.82	8.56	4.65
1980	2.71	7.69	29.11	12.05	4.50
1981	2.97	10.25	36.03	12.61	4.61
1983	4.64	10.55	40.33	14.02	5.46
1984	6.08	11.85	47.93	15.25	6.19
1985	4.15	15.78	60.46	19.27	7.05
1986	3.54	14.74	58.53	21.12	8.24
1987	4.25	13.28	55.50	22.15	9.55
1988	4.15	13.20	61.87	24.80	10.61
1989	3.79	12.95	58.02	26.29	11.52
1990	4.51	16.13	72.14	28.46	12.37
1991	5.68	16.38	84.19	31.45	13.54
Annual growth rates					
Pre-reform period					
1953–65	4.99	−0.51	5.92	4.13	3.82
1965–78	0.25	2.83	3.43	3.45	3.42
Reform period					
1978–84	18.73	14.64	12.36	10.10	4.88
1984–89	−9.02	1.79	3.89	11.51	13.23
1978–89	5.20	8.61	8.43	10.74	8.60

Notes: (a)   Sugar-bearing crops include sugarcane and sugar-beets.
  (b)   Meat includes pork, beef, mutton and other meats.
Sources: 1978 and 1981: SSB (1981a) (main indicators of the national economy); 1980, 1983, 1984 and 1985: *BR* (1985a); 1986–89, *ZGNYTJZL*, 1986, 1987, 1988, 1989 and 1991.

independence by 2.3 per cent from 1978 to 1985, mainly because of the sudden upsurge of agricultural output shortly after the reforms. Although it decreased after 1985, reflecting the long-term trend of a diminishing role played by agriculture, the reduction of agricultural share was not as rapid as in the pre-reform period. By 1989, the share of agricultural output in total agricultural and industrial output was 22.9 per cent, or 4.2 per cent lower than its peak in 1985.

Agricultural Policies and Economic Reform

Table 2.13   Structural changes in industries and agriculture

	1952	1957	1978	1980	1985	1988	1989	1990	1991
As % of total agricultural and industrial outputs									
Agriculture	56.9	43.3	24.8	27.2	27.1	24.1	22.9	24.3	22.4
Light Industry	27.8	31.2	32.4	34.3	34.6	37.3	37.7	37.4	37.9
Heavy industry	15.3	25.5	42.8	38.5	38.3	38.4	39.4	38.3	39.7
As % of total industrial outputs									
Light Industry	64.5	55.0	43.1	47.2	47.1	49.3	48.9	49.2	48.9
Heavy industry	35.5	45.0	56.9	52.8	52.9	50.7	51.1	50.6	51.1
As % of total agricultural outputs									
Crops	73.5	71.4	76.7	71.7	63.0	55.9	56.2	57.2	56.3
Livestock	11.2	12.2	15.0	18.4	22.0	27.2	27.5	26.4	26.3
Others	15.2	16.4	8.3	9.9	15.0	16.9	15.9	16.4	17.4
(a) Forestry	1.6	3.3	3.4	4.2	5.2	4.7	4.3	4.5	4.8
(b) Sidelines	12.3	11.2	3.3	4.0	6.3	6.7	6.2	6.0	5.8
(c) Fishery	1.3	1.9	1.6	1.7	3.5	5.5	5.4	5.9	6.8

Notes: Agricultural output value includes industrial output value at and below the village (chun) level.
Sources: CSYB (Chinese edition) (1990), p. 28; (1992), p. 25. ZGNYTJZL, (1991), p. 18.

Significant changes in production structure are also revealed by the structural change inside agriculture. The share of crop production in total agricultural output decreased by more than 20 per cent, from 76.7 per cent in 1978 to 56.2 per cent in 1989. The share of animal husbandry increased by 12.5 per cent from 15 to 27.5 per cent over the same period, mainly due to rapid growth in the production of meat. The shares of sideline production, forestry and fishery also increased (Table 2.13, Figure 2.2).

All these structural changes are important for agriculture and the whole economy. A higher share of light industry means more labour input and higher value-added in the industrial sector. Non-crop agricultural products are generally more labour intensive and profitable than crops. They also require less land input, which is becoming the most binding constraint on agricultural growth and rural development.

Figure 2.2   Structural changes in agricultural production

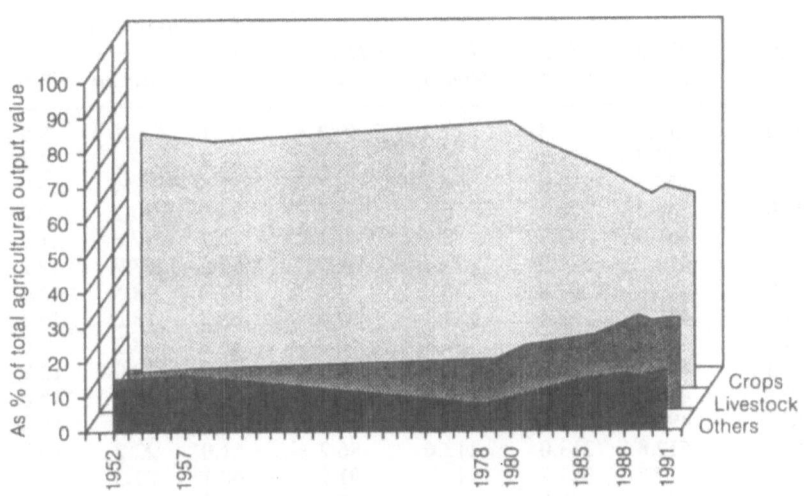

Source: Table 2.13.

Structural changes in production have been accompanied by significant changes in the structure of employment. The share of agricultural labour decreased much faster in the reform period than in the pre-reform period. It decreased from nearly 71 per cent in 1978 to 60 per cent in 1989, with an average annual decreasing rate of 1.45 per cent compared to only 0.64 per cent from 1953 to 1978 (Table 2.14).

In the rural areas, most increased labour has been absorbed by the rural non-farm sectors, such as industries, construction and services, during economic reforms. As indicated in Table 2.15, total labour force in the rural areas increased by more than 100 million people between 1978 and 1989, with an annual growth rate of 2.67 per cent. Less than half of the newly increased labour was absorbed by agriculture and the rest by rural industries and services. As a result, the share of agricultural labour force in total rural labour force decreased by more than 10 per cent from 89.7 in 1978 to 79.2 per cent in 1989. The share of non-agricultural labour force increased to more than 20 per cent of total rural labour force over the same period (Table 2.15, Figure 2.3).

As more labour is transferred from agricultural to non-agricultural activities, the share of non-agricultural labour force accounted for by the rural population as parts of the total non-agricultural labour force

Table 2.14   Structural changes of employment (million people)

Year	Total Labour	First Industry	Second Industry	Third Industry	As % of total employment First	Second	Third
1953	213.6	177.4	17.2	19.0	83.1	8.1	8.9
1957	237.7	193.1	21.4	23.2	81.2	9.0	9.8
1965	286.7	233.9	24.1	28.7	81.6	8.4	10.0
1970	344.3	278.1	35.2	31.0	80.8	10.2	9.0
1978	401.5	283.7	70.7	47.1	70.7	17.6	11.7
1979	410.2	286.9	73.4	49.9	69.9	17.9	12.2
1980	423.6	291.8	78.4	53.4	68.9	18.5	12.6
1981	437.3	298.4	81.3	57.6	68.2	18.6	13.2
1982	453.0	309.2	84.8	59.0	68.3	18.7	13.0
1983	464.4	312.2	88.1	64.1	67.2	19.0	13.8
1984	482.0	309.3	97.3	75.4	64.2	20.2	15.6
1985	498.7	311.9	105.2	81.6	62.5	21.1	16.4
1986	512.8	313.0	113.6	86.2	61.0	22.2	16.8
1987	527.8	317.2	118.7	91.9	60.1	22.5	17.4
1988	543.3	323.0	123.0	97.3	59.5	22.6	17.9
1989	553.3	332.8	121.2	99.3	60.2	21.9	18.0
1990	567.4	340.5	121.6	105.3	60.0	21.4	18.6
1991	583.6	348.8	124.7	110.2	59.8	21.4	18.9

Annual growth rates (%)
Pre-reform period

1965/53	2.4831	2.3308	2.8507	3.4969	−0.1486	0.3587	0.9893
1978/65	2.6244	1.4959	8.6311	3.8841	−1.0997	5.8531	1.2275
1978/53	2.5565	1.8958	5.8170	3.6981	−0.6443	3.1792	1.1131

Reform period

1984/78	3.0925	1.4503	5.4668	8.1579	−1.5929	2.3031	4.9135
1989/84	2.7975	1.4754	4.4908	5.6612	−1.2862	1.6472	2.7857
1989/78	2.9583	1.4617	5.0220	7.0158	−1.4536	2.0044	3.9409

Notes:
(a) The first industry includes agriculture, forestry, animal husbandry, fishery and water conservation.
(b) The second industry includes industry, construction and geographical expedition.
(c) The third industry includes transportation and communications, catering, commerce, housing, heath care and education, social services, banking and insurance, scientific research and information, administrations and others.
(d) Industry (in second industry) includes industries at and below the village level. Thus agriculture does not include any industrial activities in the villages and households. This system has been adopted since 1978.
Sources: Data are derived from the *CSYB* (Chinese edition) (1990), p. 117; (1992), p. 101.

Table 2.15   Structural changes of rural employment (million people)

Year	Total Labour	First Industry	Second Industry	Third Industry	As % of total labour		
					First	*Second*	*Third*
1978	306.40	274.90	19.60	11.90	89.72	6.40	3.88
1979	310.30	278.40	19.90	12.00	89.72	6.41	3.87
1980	318.40	283.30	22.30	12.80	88.98	7.00	4.02
1981	326.70	289.80	22.80	14.10	88.71	6.98	4.32
1982	338.70	300.60	24.50	13.60	88.75	7.23	4.02
1983	346.90	303.50	26.50	16.90	87.49	7.64	4.87
1984	359.70	300.80	33.60	25.30	83.63	9.34	7.03
1985	370.70	303.50	38.70	28.50	81.87	10.44	7.69
1986	379.90	304.70	44.50	30.70	80.21	11.71	8.08
1987	390.00	308.70	47.30	34.00	79.15	12.13	8.72
1988	400.70	314.60	49.40	36.70	78.51	12.33	9.16
1989	409.40	324.40	47.60	37.40	79.24	11.63	9.14
1990	420.10	333.36	47.52	39.22	79.35	11.31	9.34
1991	430.93	341.86	48.02	41.05	79.33	11.14	9.53
Annual growth rates							
1978/84	2.71	1.51	9.40	13.40	−1.17	6.51	10.40
1984/89	2.62	1.52	7.21	8.13	−1.07	4.47	5.37
1978/89	2.67	1.52	8.40	10.97	−1.12	5.58	8.09

Notes:
(1) The first industry includes agriculture, forestry, animal husbandry, fishery and water conservation.
(2) The second industry includes manufacturing and construction.
(3) The third industry includes transportation and communications, catering commerce, health care and education, social services, banking and insurance, scientific research and information, administration and others.
(4) Industry (in second industry) includes industries at and below the village level. Thus, agriculture does not include any industrial activities in the villages and households.
Sources: Data are derived from the *CSYB* (Chinese edition) (1990), p. 128; (1992), p. 115.

of the whole country has been increasing. The share of the *second industries* (construction and industries) employment accounted for by the rural population in total national *second industries* employment increased by almost 12 per cent from 27.7 to 39.3 per cent between 1978 and 1989. Over the same period, the share of *third industries* (services) employment accounted for by the rural population in the national total increased by 12.4 per cent from 25.3 to 37.7 per cent (Table 2.16).

Agricultural Policies and Economic Reform

Figure 2.3   Structural changes in rural employment, 1978–91

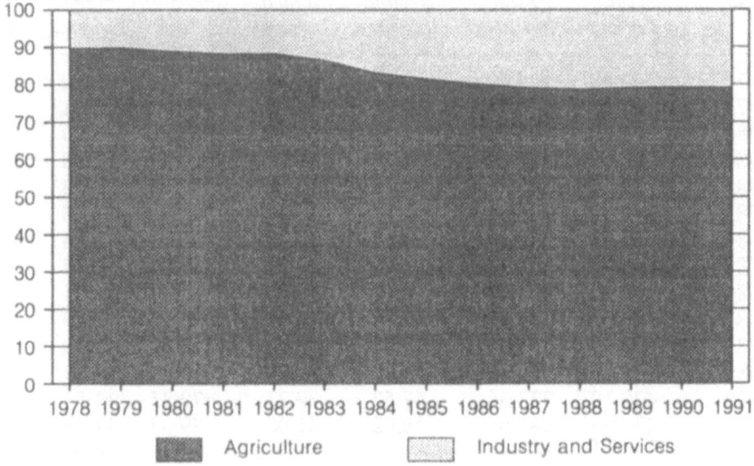

Source: Table 2.15.

Table 2.16   Rural labour as percentage of national total by sector

Year	Total Labour	First Industry	Second Industry	Third Industry
1978	76.31	96.90	27.72	25.27
1979	75.65	97.04	27.11	24.05
1980	75.17	97.09	28.44	23.97
1981	74.71	97.12	28.04	24.48
1982	74.77	97.22	28.89	23.05
1983	74.70	97.21	30.08	26.37
1984	74.63	97.25	34.53	33.55
1985	74.33	97.31	36.79	34.93
1986	74.08	97.35	39.17	35.61
1987	73.89	97.32	39.85	37.00
1988	73.75	97.40	40.16	37.72
1989	73.99	97.48	39.27	37.66
1990	74.04	97.90	39.08	37.25
1991	73.84	98.01	38.51	37.25
Annual growth rates				
1978/84	−0.37	0.06	3.73	4.84
1984/89	−0.17	0.05	2.61	2.34
1978/89	−0.28	0.05	3.22	3.70

Notes and sources: Tables 2.14 and 2.15.

These statistics suggest that increased non-farm employment has been largely due to the rural people themselves, rather than the state, which has been facing enormous difficulty in absorbing the newly increased labour force in the cities and towns. Therefore, the structural changes in employment during the reforms have been mainly attributed to the revolutionary changes in the rural areas: increased agricultural productivity enabling large numbers of labour to be freed from farming; cheap labour from the farming sector making rural non-farm production profitable and rapid expansion possible; development of non-farm production helping to increase incomes and living standards and to maintain sustainable agricultural growth.

2.3 FIELD OBSERVATIONS

For westerners or any people who are not familiar with the evolution in production and living conditions in rural China, dramatic changes during the reforms appear to be clearly illustrated by sets of statistical figures. These figures, however, may be 'suspicious': fast growth in income indices may be due to inflation (all the growth rates in this chapter are deflated) or may be manipulated by the authorities, which have a tendency to exaggerate the achievements of reforms. After presenting the impressive statistics above, it is useful to present some personal experiences of the *observed* changes in parts of rural China. Below are described the revolutionary changes during the last decade in a remote village where I lived until 1978 when I went to an agricultural university in Hainan.

The village is located 450 kilometres to the east of Guangzhou, the capital city of Guangdong. It is about 10 kilometres away from the county capital of Jieyang and 50 kilometres away from the prefectural city of Shantou, one of the four special economic zones in China.[1] Currently, there are about 2000 people in the village. In 1978, the population was about 1600. The principal agricultural product is paddy rice. Other products include vegetables, wheat, sweet potatoes, sugarcane and oranges. In total, there were 70 hectares of arable land in 1978, but the area has been declining over the last decade because of house construction and the establishment of a few village factories. Currently, there are approximately 60 hectares of land available for crop production (with only 0.03 hectares per head).

In 1978 when I left the village, most villagers were not able to overcome food shortages or buy decent clothing. Although most land was irrigated, yields of rice and other crops suffered from extreme annual fluctuations due to lack of fertilizers, pesticides and suitable rice seeds.[2]

In a normal year, after deducting agricultural taxes, seeds, waste and overheads for the three levels of management (commune, production brigade, and production team: see Chapter 1), each family member had about 10–15 kg of unprocessed rice per month (equivalent to 7–10.5 kg of processed rice). Since rice was the predominant source of income, the work points collected by the team members were given a value according to the amount of rice left for distribution. Usually one working day of an adult equivalent had ten work points, the value of which could range from 0.10 yuan to about 1.00 yuan, depending on the harvests. The team members then had to use these *booked* incomes (certainly not cash income) to buy the distributed quota of grain allocated to them. Households without *booked* incomes (households not earning work points, such as those without able adults, or those in which able adults had non-farm employment) had to pay the production teams in cash in order to buy the allocated quotas of grain.

Since the per capita entitlement of distributed grain was the residual of taxes and other predetermined obligations, farmers faced enormous problems when the harvest was poor. Even in a normal year, the entitlement of 7–10.5 kg of processed rice per month was far from meeting a *healthy* consumption level. In those years, most villagers suffered chronic undernourishment, if not malnutrition. Consumption of non-staple foods, such as cooking oil, pork and eggs, was extremely limited. More than 95 per cent of nutritional intakes had to come from rice and sweet potatoes, the two major staples for the villagers at that time. The consumption of industrial goods, such as watches, bicycles, radios and fabric clothing, was limited to a handful of households with wealthy relatives overseas.

When I paid my first visit to the village in 1980, production responsibility was slowly being introduced to the village with strong resistance from the leadership in the commune and the production brigade. Due to the changing environment and the successful outcome of reforms in the nearby communes and brigades, the leadership of the commune was forced to adopt more radical reforms in 1981 when a typhoon destroyed a large portion of the winter harvest, mainly due to the inability of the production team to harvest the rice quickly to avoid the devastating typhoon.[3] The year 1981 was one of the most disappointing in the village history since the Great Leap Forward movement. For instance, due to the poor harvest, each working day was given a booked income of 0.07 yuan. Per capita ration was less than 10 kg of unprocessed rice per month, including the state's aid (reduction of forced delivery).

In the spring of 1983, I paid my second visit to the village and conducted an informal interview with a number of old women who were drying their rice under the sun. I was told that they had two consecutive good harvests in 1982 and 1983, and due to the production responsibility system, they were able to retain a higher proportion of harvested crops. Therefore, most households started to build up their own security stocks. All the women interviewed stated that they did not have to worry about food security, at least for the coming year. I regarded this finding as the first revolutionary change in the 1980s. Farmers with little education were not aware of how political struggles had brought about policy changes, but they understood that the new policies articulated by Deng Xiaoping were much better than those pursued by Mao Zedong.

Of course, the change in 1983 was only reflected by food consumption and security. The consumption of durable goods, such as radios, good bicycles and watches was still scarce. TV sets were rapidly creeping into the urban households but they were still foreign to any villagers. However, the consumption of clothing, meat and cooking oil had started to rise. I estimate that more than 10 per cent of nutritional intakes came from non-staple food consumption in 1983 for an average household.

My third visit to the village was in the summer of 1985 when the harvest suffered a setback from its peak in 1984. To my surprise, farmers' incomes were soaring. One significant change in 1985 was that many new houses had been or were being constructed. The size of the village had almost doubled, whereas the old village had remained unchanged for centuries before 1978. Radios, watches, good bicycles and fabric clothing became very popular. The consumption of meat and cooking oil soared to a much higher level. Probably more than 20 per cent of nutritional intake was accounted for by non-staple food for an average household. Another significant change was that many villagers were engaged in non-farm activities, especially construction and trades in the cities or nearby towns. The proportion of non-farm income was probably as high as 20 or 30 per cent of total income for an average household. Most women had cash income-generating activities as well, mainly engaged in knitting work for some Hong Kong businessmen and cloth-making for local sales.

I made my fourth visit in early 1988 when the entire village seemed to be transformed from a traditional farming community to a commercial and modern village due to rising incomes and non-farm employment, although farming was still the most important activity and a major source of income. The following factors accounted for this change.

First, due to increased consumption of non-staple foods and non-food commodities, the proportion of household expenditure on the market as a proportion of total household expenditure increased rapidly. Even some of the most traditional consumption items, such as vegetables, were purchased from the market rather than grown domestically.

Second, more than 50 per cent of male adults were mainly engaged in non-farm activities inside or outside the village. The range of non-farm activities increased from construction and trades to manufacturing and transportation. Two brickmaking factories were set up in the village, employing more than 50 full-time workers, and many more on a part-time basis. Some people set up small retail and clothes-making shops in the former commune town and the county capital town. Some people went as far as Guangzhou and other provinces to look for jobs.

Third, although I found it difficult to accurately estimate the non-farm employment and income as a proportion of total employment and income, an informal survey on a number of households suggested that more than 80 per cent of cash incomes (excluding food retained for self-consumption) came from non-farm activities and at least half of total employment was not related to farming.

Fourth, at least 20 per cent of cooking fuel relied on commercial coal. In the past, rice straws were the only source of fuel for the villagers. One significant change observed in 1988 was that villagers were not short of fuel as they used to be in the past. If rain after harvest spoiled the rice straws, shortage of fuel was inevitable. The use of commercial coal to complement rice straws as fuel was made possible by increased incomes and market reforms which enabled coal to be shipped from the north to the rural south. In the past, especially before economic reforms, commercial coal was neither affordable nor available in the rural markets.

Fifth, due to the construction of more new houses, 20 per cent of the old houses in the old village were totally abandoned. For many centuries, these old houses were the only valuable property of the villagers.

Sixth, about 5 per cent of the households had colour TV sets imported from Japan. This marked a significant change in the quality of life in the village. In the past, film shows were probably the only entertainment and cultural activity. As films were shown less than once a month and the quality was poor, the appearance of TV sets in some households made an enormous change in the village because neighbours who did not have TV sets usually went to watch programmes in those who had.

The transformation has been reinforced in recent years. During my last visit early in 1992, I was amazed by the speed of change in the village. The following estimated statistics mark a number of fundamental changes in the village and they can serve as a summary of how the village has been transformed since the inception of reforms fourteen years before.

(1) The fundamental change in housing conditions

In 1988, only 20 per cent of the households abandoned their old houses. Now more than 90 per cent of the villagers do not live in the old houses. These old houses are still in good condition but they are either abandoned or used by the owners for storage purposes. The size of the new housing area which surrounds the old village is about three times as large as the size of the old village. The quality of new houses is also much higher.

(2) The fundamental change in employment

Now most male adults are not working in the field. Almost any able adult who wants to seek non-farm work will find a job in the village or the nearby towns. A typical manual worker can earn as much as 10–15 yuan per day. This is equivalent to the *booked income* of an entire month before 1978. Women who do not have much household responsibility are mainly engaged in knitting and making of clothes. On average, women are able to make 2–5 yuan per day. Some can make as much as their husbands. The major constraint on women's participation in non-farm activity is farming and household work.

(3) The fundamental change in farming

During my recent visit, I was surprised by the low level of labour input and the high level of crop yields. A large number of households had rice stocks which could easily last for more than two years without new harvests. It had been enabled by reduced delivery to the state and higher production. Most villagers preferred direct cash payment to rice delivery as taxes to the state. The market price of unprocessed rice at

0.70 yuan per kg was unbelievably low at a time when incomes had greatly increased. This price level would be the lowest market price during the 1970s; thus the real price must have been considerably lower in 1991 than in the pre-reform period.

We estimated that the level of labour input was less than 20 per cent of that during the Cultural Revolution but rice yield increased by as much as 100 per cent. A sharp reduction in labour input and an increase in yield were made possible by a combination of many factors. I tried very hard to find the possible explanations because I had not worked on the farm for more than thirteen years since I left the village.

First of all, I asked a number villagers about their recent harvests. Most of them told me that the yields from their plots were between 1200 and 1400 *jin* per *mu* of unprocessed rice.[4] This was almost twice as high as the average yield of 600–800 *jin* per *mu* during the Cultural Revolution. Of course, my first response to the answers was very doubtful. I thought that the yield at this level must have been very unusual. But after holding some more interviews with other villagers, I was convinced that the yield of that level was normal and most villagers had achieved this record during the last several harvests. This finding also helped to solve the puzzle of how villagers were able to build large private stocks of rice for their families.

As for the low level of labour input, I found it much easier to understand. During the Cultural Revolution, especially when farmers were forced to learn from Dazhai, a *model* village in Shanxi province, a lot of labour input in the field was wasted. High levels of input in the past were also necessary because a good deal of effort was made to collect different types of organic fertilizers for the crops. There was little use of chemical fertilizers which were very expensive. Unless marginal returns were high, most production teams would not use them. Sometimes when they were desperately needed, especially when crops were about to flower or near harvest, they were not available in the markets even if the teams were willing to pay a very high price. In terms of nutritional value, 1 kg of urea can be equivalent to one ton of river-mud which used to be the most important source of fertilizer in the past. To carry a five-ton boat of river-mud a distance of 10–15 kilometres to the village and apply the mud to the field, required at least thirty person days. Now farmers do not have to use such fertilizers as mud or even human and animal waste. They only use concentrated chemical fertilizers and some highly concentrated human and animal waste. The substitution of chemical for organic fertilizers has halved the total labour input used in the past.

The use of chemicals for weeding also has helped to reduce labour input significantly. Weeding in the past was done manually twice or three times for each crop season. It was not only time-consuming, but also very arduous.

Labour input has also been substantially reduced by a new pre-planting practice widely adopted since 1988. The traditional pre-planting procedure was very complicated. Seeds had to be planted in the seed beds to produce seedlings, which were then transplanted to the *main* paddy field. With the new practice, seeds are directly applied to the *main* paddy field without going through the seedling process. I saw this new practice in 1983 on a state farm in western Guangdong and tried to introduce it to my family. My father, with his rich experience in rice cultivation, said that it was a good idea but would not work in the village. After so many years, I was really surprised to see that this same practice had been widely adopted in the village and the whole region in eastern Guangdong. The new practice is not only labour-saving; it has also been one of the major factors helping to raise rice yield in the last few years. The adoption of this new practice can be regarded as one of the many revolutionary changes in recent years, mainly induced by a strong desire to save labour input, which have killed two birds with one stone (higher yields and less labour).

Excessive labour input in the past also resulted from the desire to increase total grain output by adding a winter crop (winter wheat) after the normal two rice seasons. I did not and still do not have a clear idea whether three crops could increase total output but, according to current observations, all the villagers have gone back to the two-crop patterns, one early rice and one late rice. This at least reduces labour input by another 10 or 20 per cent of total annual labour input compared with the past. In retrospect, winter wheat production in the past must have lowered soil fertility and reduced yield potential for the two major rice crops. At best, any increase in total output per hectare of arable land per year must have been overshadowed by the amount of increased labour input. At worst, increased labour input must have been entirely wasted if the output of two crops had been as high as that of three crops.

(4)   A fundamental change in living standards

Due to increased non-farm incomes and crop production, the living standards of the villagers have been greatly improved in the last few years. I did not compile statistical data because of time constraints, but the vast improvements in living standards were easily observed not only by the large number of new houses, but also by many other indicators.

As mentioned above, only 5 per cent of households in the villages had colour TV sets in 1988. By the time I was there in early 1992, at least 30 per cent of the households had their own TV sets, with many more families wanting to buy them. Many are looking for more expensive and larger ones. Surprisingly, all the TV sets in the village are from Japan. A typical Japanese 21 inch TV set costs almost one year's salary for a university professor. Although China has become one of the biggest TV producers in the world, so far villagers have refused to buy the much cheaper domestic product.

In 1988, rice straws still accounted for 80 per cent of cooking fuel for an average family. Now households do not use rice straws for cooking. These are all burned in the field to fertilize the soil. Coal has become a predominant source of fuel for all the villagers. It is affordable, and the supply has also been reliable and stable. Some villagers use coal-gas for cooking. This is a truly revolutionary contrast to the traditional village life. Before economic reforms, a typical husband would be engaged in digging river-mud to fertilize the crops and a typical housewife and her young children would be engaged in cooking rice porridge for family members and feed for the pig(s) with rice straws. Rice straws were inefficient. An average meal would take as much as one hour of the wife's time (three meals a day means three hours for cooking). The replacement of coal for rice straws probably made the most profound impact on the lifestyle of the villagers, a lifestyle which had lasted for centuries. Firstly, more than 30 per cent of the women's time is freed from cooking. Secondly, the return of rice straws to the soil will sustain soil fertility and become the foundation for increasing yield in the future. This is particularly important in the sense that the use of organic fertilizers has been declining while the use of chemical fertilizers has been increasing. Without the ashes of burned straws, soil fertility will inevitably decline.

Increased living standards are also reflected in the use of tap water, electricity and improved in-house sewage systems. Tap water has been available since 1983 but it has become much more reliable in recent years. Although electricity was first introduced to the village in 1981, the supply was utterly unreliable for many years. It was not until two years ago when one villager installed a generator that a reliable supply was guaranteed. Most families have also built their own in-house toilets during the last four years. This is a significant improvement for the environment in the village. In the past, people used open toilets outside their houses, a practice of poor sanitation and great inconvenience, especially for children and the older people.

In summary, the overall successful development in the village in the last decade has been characterized by the ability of the villagers to develop non-farm employment opportunities, the ability to cut labour input in farming for non-farm purposes and the ability to increase crop productivity. This contrasts sharply with the past when all farmers were tied to the field, digging river-mud, weeding, planting and harvesting. Off-farm activities were limited and, if available, were prohibited.

Year after year, generation after generation, villagers worried about hunger, about their ability to have daughters-in-law, to have offspring who could inherit the old houses passed on to them by their forefathers. Now the same people are abandoning their inheritance. They do not like the old houses because they are too old and too small. They do not like domestically-made TV sets because their quality is lower than the Japanese products. They do not like the outdoor toilets because they are filthy and inconvenient. They use coal instead of straws because coal can save labour and it is clean to use, and because the return of straws to the field can increase fertility. They do not need to eat too much rice because fish, meat and cooking oil now make up a significant proportion in their diet. Sweet potatoes were inferior goods in the past. The villagers liked to have sweet potatoes because they did not have enough rice. Today, ironically, sweet potatoes have become a *luxury good* because, according to the villagers, sweet potatoes have a different function: they add more fibre to the diet!

3 Policy Changes during Economic Reforms

3.1 INTRODUCTION

Following the description of the outcomes of economic reforms, the aim of this chapter is to explain why reforms have been so successful when compared to the unsatisfactory performance of economic growth and agricultural production in the pre-reform era. As was also observed in Chapter 2, the development of agricultural production, especially grain production, has not been smooth throughout the reform period. Agricultural production, and grain output in particular, suffered a setback between 1985 and 1988. It appeared to have made a strong recovery in the following three years. All this suggests that there are still many problems to be solved in the reform process, even though the reforms so far have been very successful in the sense that production growth has been much faster, per capita incomes have been greatly improved and the structures of the economy and employment have been greatly transformed.

Chinese agriculture has always been dominated by grain production. China is such a large country that food production is the foundation for sustainable economic growth and rural development. The pursuit of food self-sufficiency is desirable not only because of foreign exchange constraints on food imports, but also because of the existence of comparative advantage for food production in China if there were no market distortions. Fluctuations of food production in the past had been largely characterized by the degree of state intervention in food marketing and pricing. Even in the reform period, when market regulations and pricing were greatly relaxed, market distortion still remained the dominant factor affecting food production. Of course, there are other factors which have affected and continue to affect the incentives for food production in the future. These include low levels of state investment in agriculture and probably a long-term and inevitable trend of reduced comparative advantage in food production. To develop a clearer strategy on how to stimulate food production without discouraging the production of other agricultural commodities, it is important to conduct a detailed analysis of the new policies for agricultural growth and food production and to identify what factors have caused the fluctuations of growth since the reforms.

After the death of Mao Zedong in 1976, the whole economy (including agriculture) was in a state of chronic stagnation. There was enormous pressure by the majority of farmers on the government for reform. However, due to tremendous ideological and social resistance, real reforms could not begin until the end of 1978 when the Third Plenum of the Eleventh Conference of the Central Committee of the Chinese Communist Party was held.

Economic reforms in China were initially carried out in the farming sector with the introduction of various forms of responsibility systems in 1979. The heart of reform is to replace bureaucratic commandism with indirect planning and to give the peasantry more freedom in production and marketing. The gradual abolition of the People's Communes (which were replaced by the present economic organizations, called Xian districts), has helped greatly to reduce the undue burden on the peasantry and to eliminate inefficiency caused by the irrational behaviour of local bureaucracy. This vastly increases the initiatives and incentives of individuals.

Another feature of the reform is to encourage specialized production through market and price liberalization in agriculture. Agricultural reforms have also been accompanied by a parallel effort to balance the structure of industrial production by stopping the undue long-term emphasis on heavy industry at the expense of other sectors of the economy.

Rapid agricultural reforms have been accompanied and accelerated by the unprecedented development of rural industries and other non-farm activities, or the Rural Township and Village Enterprises (RTVEs, or the *xian zheng qi ye*). The development of RTVEs has not only eased the employment pressure in the rural areas but also accelerated agricultural growth through feedback investments from their profits.

Dramatic improvements in agricultural productivity and living standards of the peasantry have also benefited from population control. Although tough family planning has met considerable resistance, and it may create some undesirable social consequences in the long term, a sharp reduction in population growth during the reform has greatly contributed to an improvement in the average living standards and reduced the level of unemployment in most rural areas.

The success of agricultural reforms has been widely recognized at home and abroad. However, different authors have different explanations for their success. Some authors like Lardy (1983) and Griffin (1984) have attributed these achievements to the production responsibility systems, increased investments in agriculture, increased modern inputs, higher procurement prices for agricultural products, and free marketing.

Other authors like Raj (1983) and Ghose (1984) have argued that the success is mainly due to price factors combined with the infrastructure laid down during the Great Leap Forward and the Cultural Revolution. Some have gone further, to suggest that institutional reforms, especially the abolition of the commune system, will have a negative effect on agricultural and rural development in the long term. The relatively poor performance between 1985 and 1988, compared with the earlier years following the reforms, especially in the production of grain and cotton, seems to support such arguments. However, the later recovery of grain production in the 1989–91 period suggests that such a conclusion must be tentative.

As will be discussed in this chapter and Chapter 4, rapid agricultural growth and grain production during the reforms have been achieved not because of more investments compared to the pre-reform era. In fact, the share of state investments in agriculture, and the growth rate of material inputs during the reforms, have been much lower than in the pre-reform period. This finding contradicts some of the arguments by the first group of authors who have stressed the importance of increased investments and inputs for agriculture during the reforms. To a certain extent, it also supports the arguments of the later group of authors, in the sense that rapid agricultural growth has been enabled by the infrastructure laid down in the past. However, what is not convincing is that the responsibility system may become an impediment to future agricultural development in China. In fact, it is the lack of state investment in agriculture and the reluctance or inability of the government to encourage grain production through greater price and marketing incentives that explain the marked fluctuations of grain production during the reforms.

From the discussion in this chapter and the analysis in Chapter 4, it is concluded that, although there are still a number of problems and some ambiguous or undesirable government policies, agricultural reforms in China have been a great success. The major conclusions are as follows.

(1) The agricultural production responsibility system, especially the household contract system, is the key to the success of rural reforms; the benefits of any economies of scale arising from the commune system are overshadowed by individual incentives arising from the responsibility system.

(2) Specialization in production and market liberalization characterized by price reform and freer trades are two principal factors accounting for increases in allocative efficiency and farmers' incomes.

(3) The abolition of the commune system substantially reduced the overheads in agricultural production and facilitated sustained agricultural and rural development.

(4) Although the infrastructure built during the previous two decades before reforms certainly helped the upsurge of production during economic reform, sustained agricultural growth was possible mainly because of the introduction of different reform methods which increased the production incentives to farmers.

(5) Rapid development of rural non-farm enterprises had been made possible by market reforms contributing to improvement in agricultural productivity. Rural enterprises have become the major source of farmers' incomes in many areas. They helped (and will continue to help) reduce employment pressure in rural China. In fact, sustainable agricultural growth and rapid development of rural non-farm enterprises are two interdependent miracles of the agricultural reforms in China.

(6) Population control has significantly reduced population growth and facilitated an increase in per capita incomes.

(7) A setback of grain and agricultural production after the record high performance in 1984 was due to a number of factors but inadequate state investments in agriculture and lack of price incentives were the two major factors.

The following sections will discuss a set of policy changes during the reforms to boost grain and agricultural production, especially during the first phase (1978–84). These changes included: (1) the production responsibility system and the abolition of the commune; (2) structural adjustment and price incentives for agriculture; (3) market liberalization; and (4) development of RTVEs.

3.2   RESPONSIBILITY SYSTEM AND THE COMMUNE

The introduction of the production responsibility system (PRS) was the cornerstone of economic reforms. Many studies have indicated that the PRS was the most effective reform method for stimulating grain production and agricultural growth. Before the reforms, farmers were organized into production teams which, in turn, were under the control of the production brigade and the commune. This formed the so-called three-level management of the People's Commune. As discussed in Chapter 1, there were two major problems under the commune system which considerably suppressed the incentives and initiatives of individual members, or even the commune as a collective. The first was that remuneration was not related to the efforts of individuals. The second was the undue overheads imposed by the three levels of management (commune, production brigade and production team), and the irrational behaviour of commune leadership.

The PRS has gone through several stages of development. The first stage was the production contract responsibility system in which farmers were given a quota of land and other inputs. In return, they could retain all the output after delivering a fixed amount of output to the production team specified by the contract. This system applied only to the able adult members of the team and the contract had to be renewed every year. The adoption of this simple form of PRS greatly increased the team members' incentive to produce but it later transpired that there were other forms which could provide greater incentives and simplify the procedure of contract. After a few years of experiments, farmers in most areas preferred to have the so-called 'all-inclusive contract to the households' (*baogan daohu*), under which collective land is divided to individual households based on the number of able adults and the total number of household members. Under *baogan daohu*, each family is allocated a plot of land which can be used by the same family for up to thirty years, or fifty years in the more remote areas. The authorities hope that such a length of time will provide sufficient incentive for farmers to invest in land. Farmers are not given entitlement to the land because of the fear that some farmers may be forced to sell their plots and become landless, and the fear that land may be misused, especially for house construction and other non-agricultural purposes.

The adoption of different forms of PRS faced enormous resistance from the leadership in the earlier years of reforms. Due to the success of experiments in neighbouring regions, such resistance was overcome by enthusiasm from the farmers. By 1983, more than 95 per cent of production teams in the country had adopted the latest form of PRS, i.e., the household contract responsibility system, or *baogan daohu*. This remains the predominant system for the whole country up to the time of writing (1994).

The PRS has been widely regarded as the single most important reform method helping to stimulate the production of grain and other agricultural commodities. According to recent studies by a number of authors,[1] between 26.6 per cent and 78 per cent of agricultural growth during the reform period can be attributed to the incentives introduced by the PRS.

The introduction of the PRS also made inevitable the total abolishment of the commune system. The abolition of the commune greatly reduced production overheads and set a foundation for sustainable agricultural growth and rural development. Without this, many other reform methods (as discussed below) would have been far more difficult, if not impossible, to implement.

Following the introduction of the PRS in 1978, it became more and more evident that the existence of People's Communes was a tremendous obstacle to agricultural growth. On 25 August 1978, the New Chinese News Agency (NCNA) carried an article entitled 'Reform Rural Economic Management and Do Everything to Lessen the Burdens of the Peasants' by Liu Yu-cai, secretary of the Hsianyang Prefectural Party Committee in Henan province. An editor's note asserted: 'As Liu Yu-cai says, the rural collective economy keeps too much for itself and thus imposes too onerous a burden for peasants to bear. This is a long standing and unresolved problem. The implementation of the system of responsibility in joint production is so widespread in the rural areas that reform measures must be taken to cope with the situation. Under present conditions it would be practical to solve the onerous burdens on the peasants by simplifying the administrative structure, reducing the number of non-productive members, controlling unproductive overhead expenses and at the same time, strengthening and improving the system of responsibility as the Liuho Brigade has done.'

Liu's article may be summarized as follows:

(1) Since the Great Leap Forward and the People's Commune movement, peasants had been overburdened with taxes and overheads. Various forms of non-productive expenses and labour had fallen on peasants, especially during the Cultural Revolution.

Even worse, by trying to follow Dazhai, many production teams embarked upon resource-exhausting projects. As a result, less than 50 per cent of the output value of some production teams was distributed among team members. Although peasants worked hard all year round, their individual rations were less than one *jin* (0.5kg) of grain a day. In those days, it often happened that half the production teams in Hsianyang region, historically rich in fish and rice, had no cash for distribution.

(2)   Keeping too much output in reserve for communes imposed heavy burdens on the peasants. This was because there were too many cadres and members entitled to excessive subsidies and thus there were too many overhead expenses. All this was due to an improper economic system and an unreasonable structure, a legacy of long-standing leftist errors.

(3)   A basic solution for relieving peasants of this burden was to reform rural economic management, to abolish the People's Communes, and to restore the district–village administrative system. The district was an outgrowth of the county administration, the village was a basic administrative body. Village cadres were directly elected by commune members and paid by the state. It was plausible that a production team could be changed into an agricultural co-operative or several production teams could be incorporated into an agricultural co-operative in which various forms of responsibility system in joint production could operate independently under the guidance of the state plan and in which districts and villages could set up a network of economic and technical organization compatible with the productivity level, to handle economic and technical contracts.

However, as with the implementation of production responsibility system, the decision made by the reformists to abolish the People's Communes met strong resistance within the Party. The conservatives wanted (1) to persist in the principle of a 'three-level ownership with the production team as the basic unit', because a further division would deviate from the 'socialist' road; (2) to make no changes in the supremacy of integrating administration with management inherent in the People's Communes, and (3) to oppose the excessive autonomy enjoyed by production teams because this was non-socialist and wrong. The conservative view was basically Maoist, representing Mao's influence on agriculture after his death. This probably was why Deng Xiaoping and his followers had to reform the economy step-by-step.

But the reformists had absolute advantage in implementing their policies because the majority of the grass-root Chinese, especially the poverty-stricken peasants, had long suffered from the People's Communes. The problem was not that the peasants did not want to change the system; it was that they had no power to do so. The main reasons why they wanted to abolish the People's Communes, as given in Liu's article, were that it involved too many commune cadres and meetings, had too small a pie for too many persons, too many capital construction projects, too many restrictions, too many subsidies and excessive output reserves for the commune.

Liu cited the Liuho Production Brigade of the Tanho People's Commune in Hsianyang county as an example. In 1979 it had 146 brigade cadres, teachers, medical personnel and technicians in industry and sideline occupations. They earned 537,000 work points and 0.8 yuan for each labour day. The brigade paid them a total of 43,967 yuan and each brigade member received 32.95 yuan. In addition, they were provided with 93,993kg of grain of which each brigade member received 72kg. Various cadres' subsidies and free labour outside the commune added another burden to the peasants. The same condition also prevailed in other brigades of the Tanho People's Commune. The total output reserves for the commune, inflated by many subsidies and non-productive overheads generally, accounted for 20–25 per cent of the total output or even 35–40 per cent in a few brigades. The amount of grain some peasant households handed to the commune exceeded 500kg, worth several hundred yuan; in some cases the amount of grain handed over to the commune by the production teams or households accounted for more than 40 per cent of their total annual output value.

In Shangdong Province, Chuye county had 33,589 cadres at both brigade and team levels. The Hsuefuchi Commune alone had 1854 and provided 320,000 yuan for offsetting work point losses because of cadres' jobs every year. It is estimated that there were 77.2 million cadres at the brigade-team level throughout China and 16.64 billion yuan offsetting their work-point losses in 1979. This clearly suggests that the number of commune cadres was alarmingly large and that the burden on the peasants was intolerably heavy.

CCPCC Document no. 37, issued in June 1978, says; 'In 1977 production teams in Hungtang Commune of Hsianghiang county received extortionate demands from 49 agencies at county, district, commune and brigade levels. The burden on the production teams and their members could be categorized into 24 groups and 72 items, costing them 685,000 work days and 418,330 yuan in cash. Each team member had a share of 26 yuan, accounting for 42.6 per cent of his or her total annual income.' Only by abolishing the commune system could the peasants' burden be alleviated.

Advocacy of abolition was also motivated by the pattern of 'responsibility system' which was incompatible with the commune system. The household production system by contract had been widely accepted in many parts of China since 1981. The relationship between peasants and the production team had degenerated into a production contract or one of tenancy. Since peasants had assumed full responsibility for production, commune cadres were losing their command authority over peasants who were unwilling to obey their cadres any longer. This made unnecessary a great number of cadres in the production teams. Actually, most of the team cadres had to raise a contract with the team to make their own living. If the implementation of the responsibility system was to solve the rural remuneration problem, the abolition of People's Communes was needed to guarantee the system, reduce the undue burden on the peasantry, and eradicate the uneconomic and irrational behaviour of the bureaucracy.

3.3 STRUCTURAL ADJUSTMENT AND PRICE INCENTIVES

Structural adjustment in the earlier years of economic reforms was characterized by the slowing down of development in the heavy industrial sector in favour of other sectors of the economy, especially light industry.

The share of agricultural investment in total state investment increased from 10.7 per cent in 1978 to 14.0 per cent in 1979. The planned share of agricultural investment was 18.0 per cent over the period 1982 to 1984, but was curtailed to only 6 per cent on implementation. The reduction in agricultural investment was mainly due to the unusually rapid growth in agricultural production, which led the leadership to believe that 'policy and science' alone would boost sustainable agricultural growth without large investment.[2]

Although this has proved to be one of the major mistakes made during the reforms, the tremendous potential which remained untapped during the Cultural Revolution appears to have enabled the agricultural sector to grow at an unprecedented pace between 1978 and 1984 (Chapter 2, Table 2.10).[3]

Investment adjustment was mainly made in the industrial sector. The share of state investment in heavy industry was reduced from 45 per cent during 1976–80 to less than 40 per cent during 1981–85. Although it increased after 1986, in general the share of state investment in heavy industry was substantially lower in the 1980s than in the previous two decades (Table 3.1). More capital resources were allocated to the more efficient and more labour-intensive light industry, which in turn depended largely on agriculture for its raw materials such as cotton, sugar-bearing crops and oil-bearing crops. On average, the share of state investment in light industry was about 2 per cent higher in the 1980s than in the previous two decades.

Table 3.1   Basic construction investment by sector (100 million yuan)

Period	Total investment	Investment shares by sector (%)			
		Agriculture	*Light Industry*	*Heavy Industry*	*Other sectors*
1953–57	588.5	7.1	6.4	36.2	50.3
1958–62	1206.1	11.3	6.4	54.0	28.3
1963–65	421.9	17.6	3.9	45.9	32.6
1966–70	976.0	10.7	4.4	51.1	33.8
1971–75	1764.0	9.8	5.8	49.6	34.8
1976–80	2342.2	10.5	6.7	45.9	36.9
1981	442.9	6.6	9.8	39.0	44.6
1982	555.5	6.1	8.4	38.5	47.0
1983	594.1	6.0	6.5	41.0	46.5
1984	743.2	5.0	5.7	40.3	49.0
1985	1074.4	3.3	5.9	35.7	55.1
1986	1176.1	3.0	7.0	38.2	51.8
1987	1343.1	3.1	7.4	43.5	46.0
1988	1525.8	3.0	7.4	44.8	44.8
1989	1551.7	3.3	7.9	45.1	43.7
1990	1703.8	3.9	7.1	48.8	40.2
1991	2115.8	4.0	7.2	47.0	41.8

Notes:
(1) Investment made by the state for basic construction. It excludes investments by the collective and private sectors. This underestimates total investment in agriculture during the reform period because private and collective investments are more important after the reforms.
(2) 'Other sectors' includes all the sectors apart from agriculture and industry.
Sources: CSYB (Chinese edition) (1990), p. 166; (1992), pp. 158–9.

The adjustment of the national economy changed the basic structure from undue development of heavy industry to more concurrent development of agriculture, light industry and heavy industry (see Chapter 2). Higher growth of agriculture (mainly due to incentives), light industry and the service sectors (partly due to more investment) greatly helped in solving the unemployment problem during the reforms. A more balanced structure of industry and agriculture also set a precondition for improving allocative efficiency and sustaining higher growth rates of national incomes. Accumulated economic and political causes had led to a peak in the number of jobless people in the late 1970s and early 1980s. In 1979 the unemployment rate was 5.9 per cent; by 1984 it had been lowered to 1.9 per cent. From 1979 to 1984, China's cities and towns provided new jobs for 46 million people. Several million more were found each year after 1985 (*CSYB*, 1990). The employment structure also changed dramatically due to the development of rural industry. In 1979, 29.09 million (9.4 per cent of a total rural labour force of some 310 million) worked in full-time commune or brigade-run industrial enterprises alone. By 1989 the workforce of the rural township enterprises had increased up to 94 million people, or more than 20 per cent of the more than 400 million strong rural workforce (Chapter 2, and below).

Structural adjustment was accompanied by market liberalization in rural areas. As discussed in Chapter 1, the state imposed a heavy burden on the peasantry by extracting direct and indirect taxes from agriculture. A huge amount of resources was transferred from the countryside to the cities and industry in terms of cheap grain, cheap meat, cheap economic crops, and so on.

The new agricultural policy made some concessions in order to leave more resources within the agricultural sector for reinvestment. The first concession began with a 24.8 per cent increase in the price the government paid farmers for their crops, starting with the summer crops of 1979. As can be seen from Table 3.2, from 1953 to 1965 such prices had been raised fairly rapidly, with an annual growth rate about 1.7 per cent higher than the retail price index. However, during the Cultural Revolution, these rises were not large enough to make any difference, with the annual growth rate being only 1 per cent higher than the retail price index. The average increase in 1979 of 24.8 per cent can be considered all the more significant in view of the fact that even before 1963 the average annual increase was only about 3 per cent. In 1980, a 50 per cent rise in the procurement price was also introduced for above-quota cereals to be delivered to the state. This greatly raised

Table 3.2   Price index (1950 = 100)

Year	Retail price (1)	Agricultural products (2)	Rural industrial goods (3)	Worker's living expenditure (4)
1952	111.8	121.6	109.7	115.5
1953	115.6	132.5	108.2	121.4
1960	126.5	157.4	115.5	128.8
1965	134.6	187.9	118.4	139.0
1970	131.5	195.1	111.9	137.8
1971	130.5	198.3	110.2	137.7
1972	130.2	201.1	109.6	137.9
1973	131.0	202.8	109.6	138.0
1974	131.7	204.5	109,6	138.9
1975	131.9	208.7	109.6	139.5
1976	132.3	209.7	109.7	139.9
1977	135.0	209.2	109.8	143.7
1978	135.9	217.4	109.8	144.7
1979	138.6	265.5	109.8	147.4
1980	146.9	284.4	110.8	158.5
1981	150.4	301.2	111.9	162.5
1982	153.3	307.8	113.7	165.8
1983	155.6	321.3	114.8	169.1
1984	160.0	334.2	118.4	173.7
1985	174.1	362.9	122.2	194.4
1986	184.5	386.1	126.1	208.0
1987	198.0	432.4	132.2	226.3
1988	234.6	531.9	152.3	273.1
1989	276.4	611.7	180.8	317.6
1990	282.2	595.8	189.1	321.7
1991	290.4	583.9	194.8	338.1

Average annual growth rates
Pre-reform period

1965/53	1.28	2.95	0.75	1.13
1978/65	0.07	1.13	−0.58	0.31
1978/53	0.65	2.00	0.06	0.70

Reform period

1984/78	2.76	7.43	1.26	3.09
1989/84	11.55	12.85	8.84	12.83
1989/78	6.67	9.86	4.64	7.41
1991/89	2.50	−2.30	3.80	3.18

Notes
(1) Retail price index is that of all commodities.
(2) Agricultural products index is that of all agricultural products purchased by the state.
(3) Rural industrial goods index is that of industrial goods sold to the rural areas by the state.
(4) Worker living expenditure index is that in the urban areas.
Sources: CSYB (Chinese edition) (1990), p. 250; (1992), p. 236.

the income of farmers in the cereal-producing areas. A 20 per cent increase in procurement prices for live pigs delivered to the state was implemented in the same year to accelerate production of pork. Between 1978 and 1984, the price index of all agricultural goods increased at an annual rate of 7.4 per cent, or 4.5 per cent higher than the retail price index.

The second concession in the early 1980s was an expansion and reinforcement of the private plot system. With the successful introduction of co-operatives in 1956, the percentage of farmland designated as private plots stood at 5 per cent. However, in the process of the establishment of People's Communes, private plots were, in many cases, eliminated. Also, although private plots nominally remained in existence during the Cultural Revolution, one of the criteria for judging a village to be 'advanced' was the working of such land by collective labour. During the 'revolutionary' periods farming of private plots had been strongly discouraged but, with the emergence of reforming elements, the situation was reversed.

Under this new agricultural policy, the acreage under private plots increased to 10 per cent of the total. Furthermore, the limit was raised to 15 per cent in the case of areas with low productivity (ethnic minority areas, border areas, etc.).

The expansion of private plots inevitably led to the appearance of free markets in the rural communities, including those for trade in cereals. Since prices in the free markets were higher than official prices, there was further incentive for farmers to participate in them. Also, it should be noted in this connection that the increase in the acreage of private plots had stimulated the growth in side-line production.

The incentive mentioned above (that is, the rise in prices at which the government bought farm products) had benefited those farmers who were able to sell a portion of their agricultural production. Considering, however, the fact that marketed grains accounted for only 20 per cent (this increased to more than 30 per cent after 1983: Table 3.3, Figure 3.1) of the total production on a nationwide basis, it was clear that the majority of farmers were still not able to sell much of their production, and expansion of private plots had enabled smaller farmers to receive a more favourable distribution.

The third incentive was a reduction of, and even exemption from, the industrial and turnover tax in rural areas and delivery quotas in certain cases. A provisional bill regarding enterprises operated by People's Communes and production brigades promulgated in July 1979 provided for such reductions and exemptions for those enterprises.

Table 3.3   Foodgrain marketing, 1952–91 (million tons)

Year	Tax and procurement (1)	Total marketings (2)	As % of output (3)	Resales To rural areas (4)	To peasants (5)
1953	41.50	47.46	28.40	17.10	9.30
1954	45.10	51.81	30.60	23.20	—
1955	43.00	50.75	27.60	18.20	—
1956	41.70	45.44	23.60	23.00	—
1957	39.80	48.04	24.60	—	11.70
1958	55.70	58.76	29.40	—	—
1959	55.90	67.41	39.70	—	—
1960	42.80	51.05	35.60	—	—
1961	—	40.47	27.40	—	—
1965	—	48.69	25.00	—	—
1966	44.90	51.59	24.10	—	—
1967	—	49.36	22.70	—	—
1969	—	46.68	22.10	—	—
1971	—	53.02	21.20	—	—
1975	—	60.86	21.40	—	—
1977	47.20	56.62	20.00	—	16.30
1978	46.50	61.74	20.30	—	15.30
1980	49–50	72.99	22.80	—	—
1981	54.3–55.4	78.51	24.20	—	—
1982	—	91.86	25.90	—	—
1983	—	119.86	30.90	—	—
1984	—	141.69	34.80	—	—
1985	—	115.64	30.50	—	—
1986	—	115.16	33.80	—	—
1987	—	120.92	34.40	—	—
1988	—	119.95	34.90	—	—
1989	—	121.38	34.40	—	—
1990	—	139.95	36.60	—	—
1991	—	136.36	36.60	—	—

Notes
(1) Column 2 refers to total social purchases.
(2) Quantity of grain marketed was measured as 'unprocessed form' for rice and millet before 1985, as 'processed form' for rice and millet after 1985.
Sources: Column 1 – Sha Chian-li(1959); DOSW (1957b,31); Policy research Office of MOA (1980), 30. Columns 2 and 3 – *CSYB* (Chinese Edition) (1990), p. 620; (1992), p. 603. Column 4 – DOSW (1957b), 32. Column 5 – Chang Lifen (1980), 30, Agriculture Policy Research Office of MOA.

Figure 3.1   Share of marketed grain in total grain production

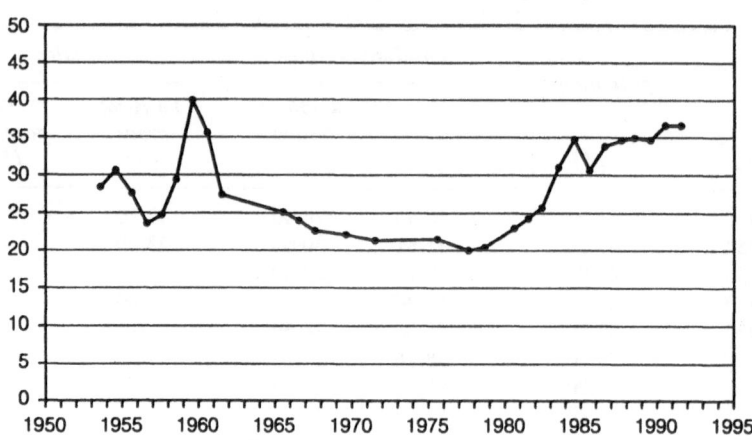

Source: Table 3.3.

Furthermore, existing small iron ore mines, coal mines, power plants and cement factories were exempted from the industrial and turnover tax for a period of three years from 1978, and new enterprises in ethnic minority areas and border areas were exempted for a period of five years.

As for the reduction of delivery quotas, this concession was granted to areas producing rice and other major cereals where per capita production of grain was 200kg or less and to areas producing other minor cereals where per capita production was 150kg or less.

The state not only raised the purchasing prices of agricultural produce and reduced the delivery quotas in some poor areas, it also subsidized the production of agricultural production materials.

Subsidization had been implemented since the 1960s. For instance, lower electricity rates were applied for agricultural use than for urban industry and general consumption. It appeared that since 1979 there had been a widening scope of agricultural production materials qualifying for subsidies or price reduction. In the early years of reforms the subsidization programmes included petroleum products and electric power for agricultural use, chemical fertilizers, agricultural chemicals and machinery. Over the three-year period from 1979 to 1981, government expenditure on subsidization for agriculture goods amounted to 34.1 billion yuan (Kojima, 1985).

Prior to economic reform, the government used a proportion of the large surplus from domestic sales of imports for production and fostering of domestic industries such as the chemical fertilizer industry. Although imported fertilizers were cheaper, the supply prices of such fertilizers were set high so that the government could pocket the difference and use it to subsidize small local chemical plants. Such local plants (and others with large equipment costs, such as cement plants and steelworks) needed subsidies from the government or local People's Communes in order to cover their deficits. Other items were also used to contribute to the government's financial revenues in much the same way. Thus the lowering of prices of such imports can be interpreted either as being a facet of the new agricultural policies, or as a policy that considerably restricted the methods by which the government had been making large profits.

3.4 MARKET LIBERALIZATION

Market liberalization during the reforms has been characterized by an emphasis on specialization in agricultural production. During the early years of reforms, the state raised procurement prices for all agricultural goods, especially grains and promoted private plot production; it also encouraged localities to specialize according to comparative advantage. Specialization, therefore, greatly improved the efficient utilization of resources within the agricultural sector.

During the period of the Great Leap Forward and Cultural Revolution, the state unduly emphasized grain self-sufficiency in all localities. The policy was 'grain is the key link in agriculture' and it totally ignored the production of economic crops as well as animal husbandry. By 1978, China not only had become a big grain importer, but also a big net importer of cotton, sugar and soybeans.

After 1978, the policy changed: 'while grain is still a key link, all methods must be taken to accelerate the production of economic crops, side-line production and rural industry'. The strict policy of grain self-sufficiency for all localities was steadily abandoned. Several steps were taken to encourage specialized production according to comparative advantage in different regions.

Beginning in 1979, the state established 13 grain-producing bases in Heilongjiang, Jilin, Sichuan, Hubei, Anhui, Jiangsu, Zhejiang, Hunan and Guangdong provinces. The land productivity of these bases tended to be more than twice that of the national average. While the national average share of marketed surplus of grain was less than 25 per cent, the average share in these bases was more than 30 or 40 per cent. The high productivity of land and marketed rate of these bases made it possible for the state to purchase a large amount of commercial grain to support the urban population and some other rural areas specializing in the production of livestock and economic crops.

Specialization during economic reforms had been characterized by two distinct changes in the spatial patterns of cereal production. One change had been the consistent reduction of areas under grains across the country. The other change was the uneven adjustment of grain–cash crop balance in various regions. These changes are illustrated in Table 3.4. The total area under grain crops increased by 0.01 per cent per year in the period from 1965 to 1978, but decreased by 0.56 per cent per year between 1978 and 1989.

In the pre-reform era, particularly between 1965 and 1978, two types of region were forced to increase grain production without comparative advantage. The first group of regions included those with good arable land for foodgrains but with greater advantage in the production of cash crops than foodgrains. These were represented by Beijing, Tijian, Hebei, Shanghai, Fujian, Guangxi, Sichuan and Guangdong, all of which had significant and positive growth in areas sown to grain crops during 1965–78 (Table 3.4). The second category of regions included those without good arable land for foodgrains but with advantages in the production of livestock products or with a need for more conservation of the environment by the growing of more trees and grasses. These were represented by Guizhou, Tibet, Yunnan, Qinhai, Ninxia and Xinjiang, all of which had the smallest reduction or positive growth in areas sown to foodgrains during 1965–78.

The changes of areas under foodgrains across regions during the reform period had exactly the opposite pattern to those from 1965 to 1978. As expected, the three metropolitan cities, Beijing, Shanghai and Tinjian, were allowed to reduce substantially areas sown with grain crops. Areas sown to grains between 1978 and 1989 decreased by 1.35 per cent per annum for Beijing, 2.5 for Tinjian and 2.19 for Shanghai. Considerable reduction of areas sown to grains had also been observed in Hebei (−1.44 per cent per annum), Shanxi (−1.01), Zhejiang (−0.64), Fujian (−0.72), Shangdong (−0.82), Guangdong (−1.47), Guangxi (−0.83), Sichuan (−0.86), Guizhou (−0.80), Tibet (−0.73), Qinghai (−0.85) and Xinjiang (−2.14 per cent per annum).

Significant reduction in areas under foodgrains had been accompanied by a much greater increase in yields, resulting in a higher growth in total output. This indicated a shift of spatial specialization helping to increase grain productivity, given that other conditions were

Table 3.4 Regional grain areas, yields and outputs

(a) Area

Regions	Area (1000 hectares) Nat. total (mil. h)				Growth rate (%, pa)	
	1965	1978	1989	1991	1965–78	1978–89
Beijing	493	561	483	483	0.57	−1.35
Tianjin	553	600	454	458	0.35	−2.50
Hebei	7225	7925	6757	6798	0.40	−1.44
Shanxi	3858	3683	3292	3200	−0.20	−1.01
In. Mon.	4716	4103	3732	3879	−0.60	−0.86
Liaoning	3821	3324	3085	3090	−0.60	−0.68
Jilin	4023	3607	3433	3542	−0.47	−0.45
Heilongjiang	6541	7138	7272	7427	0.38	0.17
Shanghai	476	532	417	416	0.49	−2.19
Jiangsu	6546	6361	6456	6203	−0.12	0.13
Zhejiang	3040	3456	3222	3267	0.56	−0.64
Anhui	6443	6177	6193	5955	−0.18	0.02
Fujiang	1725	2210	2042	2087	1.08	−0.72
Jiangxi	3877	3825	3692	3601	−0.06	−0.32
Shangdong	9259	8817	8055	8088	−0.21	−0.82
Henan	10095	9080	9252	9041	−0.46	0.17
Hubei	5829	5335	5196	5195	−0.38	−0.24
Hunan	5524	5447	5324	5365	−0.06	−0.21
Guangdong	5225	5310	4510	3872	0.07	−1.47
Guangxi	3642	3947	3601	3568	0.35	−0.83
Hainan	—	—	—	569	—	—
Sichuan	9075	10655	9688	9940	0.70	−0.86
Guizhhou	2298	2698	2471	2625	0.70	−0.80
Yunnan	3374	3669	3522	3619	0.37	−0.37
Tibet	176	206	190	192	0.68	−0.73
Shaanxi	4709	4301	4101	4088	−0.39	−0.43
Gansu	3131	2871	2822	2840	−0.38	−0.16
Qinghai	434	434	395	402	0.00	−0.85
Ninxia	773	765	707	729	−0.05	−0.71
Xinjiang	2232	2333	1840	1778	0.19	−2.14
Total	119.1	119.4	112.2	112.3	0.01	−0.56

cont. overleaf

Table 3.4 *continued*

(b) Yield

Regions	Yield (kg/hectare)				Growth rate (%, pa)	
	1965	1978	1989	1991	1965–78	1978–89
Beijing	2415	3315	4950	5790	1.39	3.71
Tianjin	2115	1950	3975	4335	−0.35	6.69
Hebei	1335	2130	3060	3330	2.05	3.35
Shanxi	1200	1830	2835	2325	1.85	4.06
In. Mon.	810	1215	1845	2550	1.78	3.87
Liaoning	1755	3360	3135	4965	2.86	−0.63
Jilin	1305	2535	4005	5640	2.93	4.25
Heilongj	1350	2070	2295	3180	1.88	0.94
Shanghai	3690	4905	5670	5955	1.25	1.33
Jiangsu	2160	3600	5085	4890	2.25	3.19
Zhejiang	3030	4245	4890	5145	1.48	1.29
Anhui	1500	2400	3915	2940	2.06	4.55
Fujiang	2640	3285	4455	4530	0.95	2.81
Jiangxi	2070	2745	4305	4620	1.23	4.18
Shangdong	1350	2595	4035	4845	2.88	4.09
Henan	1155	2310	3495	3330	3.06	3.84
Hubei	2130	2880	4650	4470	1.32	4.45
Hunan	1995	3900	5025	5100	2.96	2.33
Guangdong	2535	3405	4395	4875	1.29	2.35
Guangxi	1830	3015	3615	3855	2.19	1.66
Hainan	—	—	—	3240	—	—
Sichuan	2265	3000	4215	4380	1.23	3.14
Guizhou	2130	2385	2715	3225	0.49	1.19
Yunnan	1740	2355	2835	3060	1.32	1.70
Tibet	1650	2505	2895	3015	1.83	1.32
Shaanxi	1290	1860	2640	2565	1.60	3.23
Gansu	1185	1710	2265	2325	1.61	2.59
Qinghai	1545	2085	2805	2850	1.31	2.73
Ninxia	1080	1530	2505	2730	1.53	4.58
Xinjiang	1185	1605	3405	3795	1.33	7.08
Total	1695	2655	3690	3930	1.97	3.07

Table 3.4 *continued*

(c) Output

Regions	Output(10,000T) Nat. total (mil. T)				Growth rate (%, pa)	
	1965	1978	1989	1991	1965–78	1978–89
Beijing	119	186	239	280	1.96	2.31
Tianjin	117	117	181	199	0.00	4.02
Hebei	965	1688	2068	2269	2.46	1.86
Shanxi	463	674	933	742	1.65	3.00
In. Mon.	382	499	689	959	1.16	2.98
Liaoning	671	1117	967	1532	2.24	−1.30
Jilin	525	915	1375	1899	2.44	3.78
Heilongjiang	883	1478	1669	2164	2.26	1.11
Shanghai	176	261	237	242	1.74	−0.89
Jiangsu	1414	2290	3283	2989	2.12	3.33
Zhejiang	921	1467	1576	1679	2.04	0.65
Anhui	967	1483	2425	1782	1.88	4.57
Fujiang	456	726	910	890	2.05	2.07
Jiangxi	803	1050	1590	1626	1.18	3.84
Shangdong	1250	2288	3250	3917	2.66	3.24
Henan	1166	2098	3234	3010	2.59	4.01
Hubei	1242	1537	2416	2244	0.93	4.20
Hunan	1102	2125	2676	2682	2.90	2.12
Guangdong	1325	1808	1982	1853	1.36	0.84
Guangxi	667	1190	1302	1341	2.55	0.82
Hainan	—	—	—	178	—	—
Sichuan	2056	3197	4084	4331	1.94	2.25
Guizhou	490	644	671	886	1.20	0.38
Yunnan	587	864	998	1093	1.69	1.32
Tibet	29	52	55	58	2.53	0.58
Shaanxi	608	800	1083	1047	1.20	2.79
Gansu	371	491	639	658	1.23	2.43
Qinghai	67	91	111	115	1.32	1.86
Ninxia	84	117	177	198	1.48	3.85
Xinjiang	265	375	626	671	1.52	4.79
Total	195	305	408	435	1.98	2.49

Sources: 1965–84, *ZGNCJJTJDQ* (1949–86), p. 174–88. 1989, *ZGNYTJZL* (1989), p. 38. 1991, *ZGNYTJZL* (1991), p. 36; *CSYB*, 1992, p. 362.

unchanged. The economic rationale will be discussed in detail in later chapters.

This new policy encouraged farmers to trade their surplus grain in the free markets after meeting the procurement quotas. Consequently, the proportion of grain marketed, including sales to the state at negotiated prices and sales on rural and urban markets, rose from 20.3 per cent in 1978 to 25.9 in 1982.

In the following years, it continued rising, to almost 35 per cent by 1989 (Table 3.3). Private transactions in markets were 5 million tons both in 1979 and 1980. State purchases of cereals at negotiated prices in rural markets rose from 0.215 million tons in 1977 to 3.25 million tons in 1978, 5.28 million tons in 1979, and 8.6 million tons in 1980 (Ministry of Agricultural Policy Research Office, 1980, p. 30; Ministry of Food Research Office, 1981, p. 13; Report 1980b). Between 1978 and 1989 the urban population grew by at least 100 million. Adding to this farmers engaged in non-agricultural production activities, the increased number of non-agricultural population must be more than 180 million. However, per capita production in 1989 was 367kg, about 16 per cent higher than in 1978 (Table 2.11). Although importation of grain rose after 1977, the increment remains modest, far from sufficient to meet the demand of the additional non-agricultural population.

In fact, with the help of the strict control of population and the rapid expansion of grain production between 1978 and 1984, China had a grain surplus in 1985 for the first time since 1961. Between 1979 and 1983, the average annual amount of grain imported was 14 million tons, but in 1985 (Table 2.11). This was a tremendous achievement for the new agricultural policy. Unfortunately, the subsequent stagnation of grain production in the following four years made China a net importer of grain after 1986. In 1987, total net imports amounted to 13 million tons. Although this figure subsequently declined, China remained a net importer of cereals even after a strong recovery of grain production between 1989 and 1991 (Table 2.11).

The sudden change in China's grain trade position during 1984–85 made many economists and policy-makers in China too optimistic about the grain production of the country. Although this will be discussed in detail in Chapter 4, it is worth recalling the analysis undertaken by Reeitsu Kojima, a Japanese economist, who predicted that China would become a large permanent net importer of grain under the new agricultural policy at a time when China emerged as a net exporter. He listed the following reasons to support his argument that China would become a large food importer with the new agricultural policies (Kojima, 1985):

(1) the policy concession that the government made to farmers in connection with delivery quotas;
(2) the increase in urban population resulting from the new agricultural policy; and
(3) the change in the structure of consumption.

The stagnation of grain production from 1985 to 1988 seems to have supported his prediction. However, the reasons he listed may not really explain why China should become a large net importer of grain under the new policy. As will be discussed in Chapter 4, lower price incentives and long-term neglect of agricultural investment have been the two major factors explaining the fluctuations in grain production and agricultural development during the reforms. Thus, increasing grain deficits in China may not be inevitable, as suggested by Kojima. The reversed trend of grain production since 1989 may have helped to answer this.

It is obvious that Kojima stressed the demand side of the grain situation in China (points 2 and 3). Increased non-farm population and the changed structure of consumption due to increasing income will both boost grain demand. On the supply side, there still exists large potential for growth in grain production. This is not stressed by Kojima and it is the most vulnerable point of the state agricultural policy: good harvests in some years lead to optimism and complacency; lack of effort thereafter leads to stagnation; reinforcement of effort leads to a strong recovery. The key question is: should this cycle be repeated? Or can it be avoided by having a more strategic and consistent policy? There is not a single answer to this question, but it will be addressed throughout the rest of the book.

Not only did the state encourage specialized production in grain-producing regions, it also encouraged specialized production of economic crops in those areas which had comparative economic advantage in the production of cotton, oil-bearing crops, sugar-bearing crops and so on. A number of cotton-producing bases in the northeast, northwest and central China were set up after 1978. A number of sugar-producing bases were also set up in Fujian, Guangdong provinces and Guangxi Autonomous Region. The structure of farming areas has been improved by allocating more areas sown to these crops through price regulations.

The share of area sown to grain as a proportion of total crop area declined by almost 4 per cent from 80.3 per cent in 1978 to 76.6 in 1989, while the shares of area sown to industrial crops increased from 9.95 per cent to 14.3 over the same period. In absolute terms, total area sown to grain crops decreased by 0.65 per cent per year, while that to industrial crops increased by 3.46 per cent per year between 1978 and 1989 (Chapter 4, Table 4.4).

Specialized production of economic crops was greatly accelerated by the reopening of rural free markets, especially the free transactions of grain in them. Although publicly marketed grain increased slowly, the quantities of grain sold in the free markets were significant. In 1979 and again in 1980 private sales were 5 million tons (NCNA, 2 March 1980; Ministry of Food Research Office 1981, p. 13). Although that is less than 2 per cent of output, it is about 10 per cent of grain collected as taxes or procured by the state as quota and over quota deliveries, and 33 per cent of resales. Although little, if any, of this grain would enter long-distance transport, increased private local marketing of grain and industrial crops and sideline products facilitated specialized production within local markets. In subsequent years, marketed grain was increasingly dominated by private trade as government restrictions on movements of grain, both intra-regionally and inter-regionally, were gradually lifted. This is signified by the increasing share of marketed grain as total production throughout the 1980s (Table 3.3).

In addition to reopening rural free markets, the state expanded the number of long-term commitments for the delivery of state-procured grain to regions of specialized agricultural production. The first example, the supply of 100,000 tons of grain annually to Fujian Province, beginning in 1976, led to a significant growth in specialized production of sugarcane. In August 1980, the state council approved a five-year commitment of 225,000 tons of state-procured grain to Hainan Island for sale to the state farms which produced tropical crops and plants (Lardy, 1983).

Specialized production of economic crops was also encouraged by the change in the agricultural taxation system in 1985. Under the new policy, farmers did not have to pay their tax in grain; they could pay in cash. Although the new policy may have been inimical to grain production as discussed in the next chapter (it is also stressed by the first point made by Kojima: see above), it offered a good opportunity for farmers to make their own production decision according to market prices and local conditions of production.

The remarkable growth in agricultural production after 1978 may be due to the potential productivity which was not fully utilized under the old policies. More modern inputs, as well as the appearance of various new varieties of crops, may also have contributed to these achievements. Of course, China has to increase agricultural production by applying more modern inputs because of the limited endowment of arable land. Between 1978 and 1984, the annual growth rate of large and medium sized tractors was 7.43 per cent, that of small and working tractors was 15.67, motors for drainage and irrigation machinery 3, chemical fertilizer applied 12.3, and rural electricity consumed 10.55 per cent (Table 3.5). Between 1984 and 1989, the growth rates of these major agricultural inputs are lower than those in the 1978–84 period except the figure for electricity used in the countryside and that for drainage machinery. All these increases, however, were much smaller than those between 1965 and 1978. During that period, the average annual growth rates of the same material inputs are respectively 17 per cent for large or medium sized tractors, 57 per cent for small tractors, 16 per cent for motors for drainage and irrigation machinery, 12.4 per cent for chemicals fertilizers and 16 per cent for electricity.

Lower growth rates of material inputs in agriculture during the reform period as opposed to the pre-reform period suggest that higher production growth during the 1980s (Chapter 2) was largely due to increased efficiency of resource use. This argument is reinforced by the fact that labour and land inputs during the whole of the 1980s had declined much more sharply than before. It has also been proven by many available quantitative analyses (see Fan, 1991; Yao and Xie, 1991).

3.5 RURAL TOWNSHIP AND VILLAGE ENTERPRISES (RTVEs)[4]

The pre-reform development strategy was narrowly focused on the formal industrial sectors in the urban areas and industrial bases run by the state, or the so-called state-owned enterprises. Rapid development of the state-owned industrial sector was essential for China to shift from an agrarian society to a more modern and industrialized country, but the progress in industrialization had been largely financed by resources being transferred from the rural and agricultural sectors. After thirty years of development from 1949 to 1978, there had been

Table 3.5 Growth of agricultural inputs, 1965–91

Items	1965	1978	1984	1989	1990	1991	Growth rate (%, p.a.)		
							1965–78	1978–84	1984–89
Large/medium size tractors ('000)	73	557	857	848	814	784	17.0	7.4	−0.2
Small and working tractors ('000)	4	1373	3289	6555	6980	7303	56.7	15.7	14.8
Drainage and irrigation machinery (1000hp)	9074	65575	78281	93930	97050	98795	16.4	3.0	3.7
Chemical fertilizers applied (100% effective, 1000t)	1942	8840	17731	23600	25903	28051	12.4	12.3	5.9)
Rural electricity consumed (bil. kwh)	3.7	25.3	46.2	79.0	84.5	96.3	15.9	10.6	11.3

Notes: hp = horse power; t = ton; kwh = kilowatt hour, 1hp = 0.7355 kwh.
Sources: 1965, 1978: SSB (1981a) 'main indicators of the national economy.'; 1984: *BR* (1985a). Annual growth rates: calculated from the *previous figures.* 1989–91: *ZGNYTJZL,* (1989), p. 316; (1991), p. 272 and p. 304.

little impact on the whole rural economy in terms of employment and income.

Economic reforms in the 1980s greatly increased farmers' incomes and agricultural productivity. These created two fundamental conditions for the development of RTVEs. First of all, increased incomes and savings have enabled farmers to set up their factories or shops with their own resources. Secondly, increased agricultural productivity has enabled the farming sector to provide a large surplus of agricultural goods, and release a large number of surplus workers for non-farm enterprises.

The development of rural non-farm activities has also been facilitated by state policies through tax exemption for up to two years after establishment and the provision of enterprise credits by the government.[5]

Like agricultural reform, however, the development of RTVEs encountered strong resistance from the leadership in the early years. There was hectic debate about the nature of these enterprises, especially those run by individuals. Opponents argued that private ownership should not be encouraged in a socialist society to avoid the exploitation of labour and increased income inequality. Thus, development of rural non-farm enterprises should be entirely restricted to collective ownership. Although collective firms are more dynamic and more responsive to market signals than state-owned enterprises, labour inertia arising from the spirit of 'eating from the same big rice pot' exists in these firms. In addition, individual farmers have little incentive to make investments in firms over which they have no direct control or influence.

As agricultural reform deepened and farmers' incomes rose, there was increasing pressure on government to encourage farmers to divert their savings for productive purposes. Coupled with gradual ideological liberalization in the central government, private enterprises were allowed to be established by individual farmers in 1984 under the condition that any single owner would not employ more than eight people. The restriction on the number of workers which could be employed by any enterprise owner was actually much more relaxed in some provinces. For example, Guangdong was granted greater autonomy to set its own regulations for the development of RTVEs. It was one of a few provinces to announce detailed regulations and measures to promote RTVEs and encourage private ownership (enterprises). In December, 1986, the Guangdong government issued a number of promoting measures, some of which are listed below.

(1)   The principle of multi-level, multi-form, and multi-ownership should be adopted to promote RTVEs. It involves the promotion of primary, secondary and tertiary industries simultaneously; the promotion of large, medium and small industries together; and the development of private as well as collective enterprises.

(2)   Funds should be allocated for the promotion of different forms of RTVE. The enterprises themselves should provide the bulk of resources required, but they are also allowed to issue shares which can be bought by workers and any people outside the firms. If bonds are issued, the interest rate is allowed be to flexible and higher than the bank rate. Second, rural credit co-operatives are given more autonomy to enable flexible administration and management of enterprises. Third, private credit and capital accumulation are encouraged. And lastly, depending on their local financial situation, the cities (prefectures) and counties set aside a certain proportion of tax revenue collected from rural enterprises as a development fund to assist development of rural enterprises in backward areas.

(3)   Newly-established enterprises are granted a one-year exemption from tax payments. In some poor areas, newly-established enterprises suffering from financial difficulties may be allowed tax exemption for the second year.

(4)   Scientific and technical teams within enterprises should be strengthened to promote technological improvements. Science and technology departments and institutions must encourage the diffusion of their research results to RTVEs. The departments in charge of RTVEs must establish appropriate training pro-grammes to train technical and managerial personnel. Graduates from universities and technical colleges and state employees with specific skills will be offered incentives and other privileges by the state to work in RTVEs (Guan Ze-wen, 1991, pp. 65–7).

Similar policies were also adopted in Zhejiang, Jiangsu, Shanghai and Fujian in the same year, and in many other provinces and cities in 1987 and 1988. The legalization and encouragement of private enterprises led to an unprecedented pace of development of RTVEs in the late 1980s.

The gross output of RTVEs increased from 49.3 billion yuan in 1978 to 67 billion in 1980. It then soared to 275.3 billion yuan in 1985, and 958.1 billion yuan in 1990. Between 1978 and 1985, the real annual growth rate of the gross output of RTVEs was about 28 per cent. From

1985 to 1989, it was over 32 per cent. By 1991, non-farm enterprises accounted for 25 per cent of the total national output (7.2 in 1978), 64 per cent of total rural output (24 in 1978), 33 per cent of total national industrial output (9.2 in 1978), and 25 per cent of total national exports (4.8 in 1985). The number of workers employed by these enterprises increased from 28 million in 1978 to 96 million in 1991, or from 9 to 23 per cent of the total rural labour force (Figure 3.2). Between 1986 and 1990, non-farm enterprises contributed 54 billion yuan of their profit to support agricultural production, rural infrastructure investment, rural education and many other social services. This was more than twice as much as the total state investment in agriculture (about 23 billion yuan) over the same period (Tables 3.1 and 3.6; the Department of Policy and Law, Ministry of Agriculture, China, 1989, p. 286–7; Ministry of Agriculture, China, 1989, p. 192).

In short, non-farm enterprises have become increasingly important in all aspects of the rural economy in China, including agricultural production, infrastructural investment, employment and the improvement of people's living standards in the rural areas. On the other hand, rapid development of these enterprises has been possible because of the successful performance of agriculture during the reforms. These farm and non-farm linkages of the rural economy will become increasingly important as economic reforms deepen. This has been recognized by many analysts both at home and abroad.

The development of non-farm enterprises and their impact on agriculture have been actively studied in recent years (e.g. Economic and Policy Research Centre, MoA, 1991; Byrd and Lin, 1990). However, most available studies are very descriptive and partial. There are still many questions needing to be answered relating to the linkages of these two sectors of the rural economy.

For instance, what are the principal factors affecting the rate of labour transfer from the farm to non-farm enterprises? Can the government help to facilitate this process and how? As state investment in agriculture has been shrinking during economic reforms, the feedback investment from non-farm enterprises to agriculture will become more and more important in sustaining agricultural growth. However, how much can this feedback investment help to stimulate agricultural growth? What are the implications for the development of non-farm enterprises themselves if they have to bear an increasing burden of agricultural investment? What can the government do to reduce such a burden? What are the major factors affecting regional disparities in the development of non-farm enterprises? And what are

Table 3.6 Contribution of the RTVEs to the rural economy

Year	Values							Growth rate (%, p.a.)		
	1978	1980	1985	1988	1989	1990	1991	1978–85	1985–89	1978–91
No. of RTVEs (million)	1.6	1.4	12.2	18.9	18.7	18.5	19.1	34.3	11.2	21.3
Output of RTVEs (bil. yuan)	49.3	67.0	275.3	701.8	840.3	958.1	1161.2	27.9	32.2	27.5
No. of employees (million)	28.3	30.0	69.8	95.5	93.7	92.6	96.1	13.8	7.6	9.9
As % of total rural output	24.2	24.0	43.4	58.1	58.0	59.6	61.1	8.7	7.5	7.4
As % of total rural employment	9.2	9.4	18.8	23.8	22.9	22.6	22.3	10.7	5.1	7.0
As % of farmers' net income	7.0	10.1	24.6	—	—	—	—	19.7	—	—
As % of total national income	—	—	4.8	15.2	—	—	—	—	—	—
Total taxes and profit (bil. yuan)	11.0	14.4	42.5	98.2	96.8	101.2	118.8	21.3	22.9	20.1
(a) Taxes delivered to state	2.2	2.6	13.7	31.0	36.5	39.2	45.5	29.9	27.8	26.2
As % of total state tax revenue	4.2	4.5	6.7	13.0	13.4	—	—	6.9	18.9	—
(b) Profit retained	8.8	11.8	28.8	67.2	60.3	62.1	73.4	18.4	20.3	17.7
(i) Subsidies to agricultural production (bil. yuan)	2.6	2.3	3.0	5.5	7.1	7.8	8.7	1.9	23.9	9.6
(ii) Investments in rural infrastructure and social services	3.9	4.9	8.3	10.9	9.1	10.5	12.2	11.4	2.4	9.1
Total fixed assets (bil. yuan)	23.0	32.6	91.0	210.0	250.0	285.7	338.5	21.7	28.7	23.0
Total wages (bil. yuan)	8.7	11.9	47.2	96.3	105.5	113.0	130.5	27.4	22.3	23.2
Productivity (yuan/worker)	1744	2232	3944	7352	8971	10893	12691	12.4	22.8	16.5
Tax and profit/100 yuan of fixed assets	48	44	47	43	39	35	36	-0.4	-4.6	-2.1

Notes: (a) All the values in this table are at current prices.
(b) The total tax and profit includes 10% of pre–tax profit used for rural investments.
(c) Investments for rural infrastructure and social services include only those from the profits after tax.

Sources: ZGNYTJZL (1989), p. 192; ZGNCJJTJDQ (1949–86), p. 286–7; ZGNYTJZL (1991), p. 190.

Figure 3.2   Contribution of RTVEs to rural production and employment, 1978–91

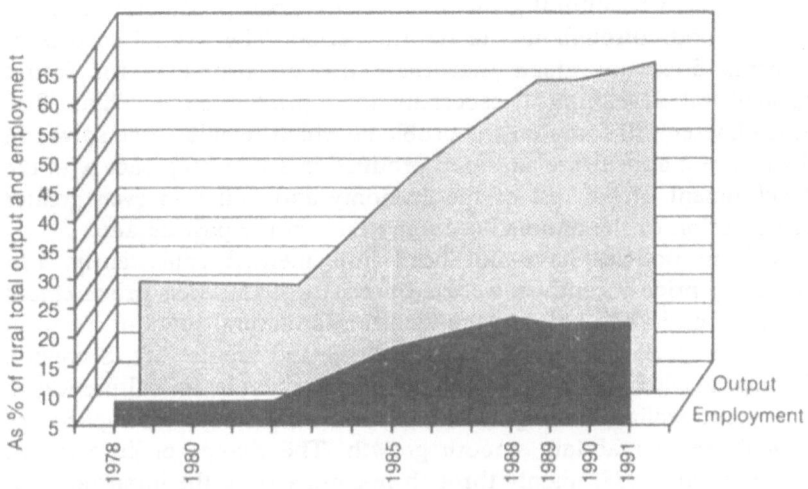

Source: Table 3.6.

the most effective means by which the development of non-farm enterprises can be accelerated in the remote and backward regions?

To answer these questions requires a systematic and quantitative analysis. However, it appears that there is an enormous gap between descriptive studies and quantitative methods in this area in China. I believe that the linkages between agriculture and RTVEs will be a very interesting area for future research.

3.6   CONCLUSIONS

Economic reforms have brought significant improvements in living standards for the people and profound structural change in production and employment, both economy-wide and within agriculture. Detailed discussion of these changes is presented in Chapter 2.

There are several inter-related factors explaining the success of economic reforms, especially in agriculture and the rural areas. These include: (a) the introduction of the PRSs and the abolition of the commune system; (b) structural adjustment and price incentives;

(c) market liberalization in agriculture; and (d) the development of RTVEs.

However, agricultural growth, especially grain production, has not been smooth throughout the reform period. The overall success of reforms does not mean that there are no problems for future agricultural development, especially food production in China. The next chapter will focus on the problems which remain to be solved in the future if agriculture and food production are to keep pace with the development of the rest of the economy and make an even greater contribution to the national economy than in the past decade.

Reform policies have not been implemented consistently. For example, price incentives were reduced in 1985 when procurement prices of grain and cotton were frozen after several years of increase, while the market prices of grains, cottons and inputs rose. Secondly, during the whole reform period, public investment in agriculture was in consistent decline, leading to a reduction in the long-term capacity of agriculture to maintain smooth growth. The change of government policy in late 1988, mainly through re-emphasizing the importance of price incentives to grain producers and retreating to a more rigid control on acreage under foodgrains, has helped to break the stagnation of grain production which prevailed between 1985 and 1988. Grain production reached new record levels after 1989.

The next chapter is therefore devoted to a more detailed discussion on why grain production has been fluctuating while agricultural production as a whole has been growing more smoothly during the whole reform period.

4 Problems and Prospects of Grain Production

4.1 INTRODUCTION

The success of economic reforms has been characterized by higher growth rates and dramatic improvements in the living standards of the population, particularly the rural population. Structural changes in production and employment are also important for long-term sustainable growth and rural development. However, the growth of different indicators and the structural changes of the economy have been by no means consistent during the reform period as demonstrated by the tables in Chapters 2 and 3.

Agriculture, grain production and farmers' incomes had a significant and steady growth between 1978 and 1984. The growth of all major indicators, especially grain output, slowed down and fluctuated considerably between 1985 and 1988. Momentum of growth recovered after 1989. Grain output reached a new record high in 1989 and 1990. Despite the widespread and devastating flooding in 1991, grain output in that year still registered the second highest ever (Chapter 2, Table 2.11).

After more than forty years of industrial development, agriculture's contribution to the national economy has declined to less than 30 per cent of GDP. The contribution of grain production to agricultural output value has also declined steadily, especially during the reform period when non-grain production is greatly encouraged. However, experience has indicated that grain production is still one of the most important indicators of national economic strength. In general, if grain output is high, market prices will be stable and the following year will see healthy economic growth. If grain output is low, market prices will go up and the whole economy will suffer the following year. This is because food consumption still accounts for more than 50 per cent of total household expenditure.[1] Food shortage requires imports and foreign exchanges. As will be discussed later, imported grains are much more expensive than those produced domestically, not only because the former require scarce foreign exchanges, but also because producer prices are much lower than import parities for all grains.

The fluctuations of grain output are paradoxical in a period of reforms and open policies. Even in the literature, scholars have had different assessments on the reforms. Almost all analysts have pointed out the importance of price and institutional reforms, especially the responsibility systems. Many scholars, however, had difficulties in interpreting the changes of price instruments and the allocation of state investments. For instance, Lardy (1983) stresses the importance of increased state investments in agriculture, but later evidence indicates that the share of agricultural investment has actually declined by almost 200 per cent during the 1980s as opposed to the pre-reform period.[2] The decline in state investment led to a deterioration of the rural infrastructure, especially the irrigation and drainage system. This, therefore, has become a focus of criticism by those who have different opinions on rural reforms (see Ghose, 1984).

This chapter attempts to summarize the major problems during the reforms and discusses the prospects of grain production in the future. After the introduction, Section 4.2 focuses on the problems and constraints on agriculture and grain production. Some problems reflect the long term state bias against agriculture, other problems have just emerged during the reforms. The problems identified include: (a) low state investments in agriculture; (b) the price and marketing structure against grain production. Two types of constraint on grain production are identified: (a) resource (especially land) constraints; and (b) economic constraints, i.e., the declining profitability of grain production.

Section 4.3 discusses the prospects of grain production in the future. Firstly, the prospects depend largely on the ability and willingness of state to reduce the constraints and solve the problems which have hindered, and will continue to hinder, grain production. Secondly, the prospects for future grain production will depend on China's comparative advantage in grain production. Empirical results suggest that China still has a significant comparative advantage in grain production (World Bank, 1991).

4.2 PROBLEMS AND CONSTRAINTS

4.2.1 Problems of the new policies

Rapid expansion in agricultural production and the rise in living standards of the peasantry do not mean that agricultural reforms have been successful in every aspect. The state has incurred substantial budgetary costs during the reforms. Increasing the procurement prices of agricultural produce has inevitably increased the prices to consumers (state employees and the whole non-agricultural population, or urban dwellers in a narrower sense). To protect the interests of the urban dwellers, the state has either paid the factory workers and state employees a certain amount of premium to cover the higher costs of agricultural products, or maintained the same consumer price at a low level (lower than the procurement price and the free market price). Therefore, the state has been buying agricultural products from farmers at a relatively high price (although it is lower than the market price) on the one hand and selling them to the consumers (urban residents) at a much lower price on the other hand. To cover the price difference, the state has incurred tremendous budgetary costs.

For instance, the state had to spend 20 per cent of its budget revenue to make up for the price differences each year between 1980 and 1985 (*The Economist*, 1 August 1987; Table 4.1). In 1984 alone, 28 billion

Table 4.1 State consumer food subsidies

Year		Consumer food subsidies	
	Value (bil. yuan)	As % of state expenditure	As % of total GDP
1979	11.5	9.0	2.9
1980	17.2	14.2	3.8
1981	25.0	22.4	5.6
1982	24.3	21.1	4.7
1983	26.3	20.3	4.5
1984	28.3	18.3	4.1
1985	26.1	14.1	3.1
1986	24.7	10.6	2.5
1987	25.9	10.6	2.3
1988	26.4	9.9	1.9

Notes: The values are in current prices. The estimates of consumer food subsidies include the government's internal figures for its losses as the sum of (a) the difference between state procurement costs and the sale revenues plus (b) state agencies's annual operating losses on the transport, handling and storage of these foods.

Source: World Bank (1991), Working Paper no. 2, vol. I, Table 4, p. 13.

yuan was spent on food subsidies. Although total subsidies have since fallen in real terms as a proportion to the budget (see Table 4.1), in absolute terms, it has been more than 25 billion yuan per year. The two-tier price system (basic procurement price and premium price) put the state in such a situation that the more it purchased grains from the farmers the more it had to pay.

Consumer subsidies in Table 4.1 do not include the net loss of farmers' revenue on forced delivery of grain and other products to the state at low prices. If this is accounted for, total consumer subsidies are likely to have been more than twice as high as they are in the table, as the estimated loss of farm revenue through forced delivery has on average been roughly the same as or even greater than the net loss to the state marketing agencies. For example, the net loss of potential farm revenue on mandatory grain sales in 1988 is estimated to have been about 23 billion yuan (World Bank, 1991, Working Paper no.2, vol. I, p. 1). Grain is not the only product which is subject to implicit state taxation; there are other products, such as cotton and oil-bearing crops. Thus the total transfers from agriculture should have been greater (more estimates will be given in Chapter 5).

The bumper harvest of grain in 1984 and heavy losses to the state marketing agencies induced the government to introduce new policy measures in 1985.[3] These included the following.

(a)   In order to strengthen the production of farm crops and increase rural incomes in general, the state tried to liberalize price control of most farm products, leaving them to find their market value. Grain as a special commodity (very important to the lives of one billion Chinese people), however, was an exception and its price remained controlled by the state.

(b)   The state introduced a single mixed grain price to replace the previous two-tier price system. The single mixed price was in between the basic procurement price and premium price.

(c)   The total grain quota of all the regions was reduced from 117 million tons in 1984 to 75 million tons in 1985 (*The Economist*, 18 January 1986).

(d)   The original compulsory procurement of grain was replaced by a contract procurement. In theory the contract was voluntary, but in reality it was not. The contract was enforced by village chiefs and local officials; farmers could not contract to supply food at all. However, although the contract was in practice the same as compulsory procurement, farmers were allowed to pay for agricultural taxes by cash rather than by grain. In addition, if farmers could not deliver as much as the contract stated, they were allowed to buy grain from the market place at a higher price and sell to the state at the mixed procurement price to fulfil the state *contract quota*. Furthermore, local governments in many regions imposed the *contract* (production quota) not through acreage, but through final output or the amount of delivery.

(e)   The prices of farm inputs were increased as part of price reform.

The first policy change actually makes grain production less profitable than other farm crops. As mentioned previously, free market prices of farm crops were usually twice as high as state procurement prices even after the latter had been significantly increased in 1978. As farmers are allowed to sell non-grain crops in the free markets while part of grain (and cotton) has to be sold to the state at a low and controlled price, production of grain inevitably becomes less profitable.

The introduction of a single mixed procurement price[4] for the two-tier price system may have discouraged the most efficient regions from supplying more grain to the state. The former premium price could be as high as (or very close to) the free market price in the more efficient (or grain surplus) regions. In that case, the premium price might become the floor price to farmers. Although the mixed procurement price is higher than the former basic procurement price, it is lower than the premium price. Therefore, farmers in these 'efficient' regions might find it unprofitable or undesirable to sell the 'above-quota' grain to the state as they did in 1984.

In those regions with relatively higher production costs or less advantage in grain production, but more advantage in non-grain crops or animal husbandry, local market prices are always much higher than state procurement prices. Therefore, the production of grain in these regions largely depends on the level of own consumption and state quota. As the quota is reduced, or the producers are allowed to pay the state in cash rather than grain for agricultural taxes, grain output is doomed to decline: these are exactly the changes of (c) and (d). Farmers would turn over rice and wheat fields to more free-market vegetables and what the Chinese called 'sideline' activities, such as poultry raising. Evidence of decreasing grain output has been widely observed in many areas in China since 1985 when the above changes were introduced. Total area sown to grains decreased by 4 million hectares or 3.5 per cent between 1984 and 1985 (Table 4.4). Grain output per hectare decreased by 135kg or 3.73 per cent. Therefore, total output decreased by more than 7 per cent in that year (Chapter 2, Table 2.11). Although total grain output recovered in the following years, by 1988 it was still lower than the peak level in 1984.

Between 1985 and 1988, grain production declined most in regions which had been experiencing rapid economic growth and fast development of the rural industries such as Guangdong, Fujian, Zhejiang and Shanghai. For example, even the major grain producing Bolo county situated in the western Pearl River Delta of Guangdong Province, experienced a significant decline in grain output in 1985 (−4.23 per cent) and 1986 (−5.3 per cent). In 1986, only 9 of the 22 townships in the county managed to deliver enough grain to fulfil the contract procurement assigned by the state. To meet the whole *contract* procurement quota, the county government had to purchase 22,000 tons (or 8.1 per cent of the total output in that county) from other counties of Guangdong or other provinces (Wei, 1987).

Stagnant grain output between 1985 and 1988 caused a series of problems. Once more, the country had to import sizeable quantities of grain to meet the expanding domestic demand. The rapidly increasing incomes of the farmers and state employees had helped increase the demand for non-grain food and meat whose production depended largely on grain as input. However, increasing demand and the restricted supply of grain not only pushed up the grain market prices but also the prices of meat and other crops. The soaring prices of agricultural produce accelerated inflation to such an extent that the government had to slow down the process of economic reform, especially reform of the price system.

There are other factors which have negative effects on grain production. More and more stories emerge not only of the failure to expand irrigation systems but of the failure to maintain those systems already in place. Irrigation ditches become clogged and pumps broken. Irrigated land in China fell by over 2 per cent from 1985 to 1987 as local government officials lost their authority to requisition labour for public works construction. While the government loudly trumpeted its *agriculture first* policy, and watched yields rise, direct state investment in agriculture fell from 12 per cent of the national budgetary expenditures before 1978 to less than 5 per cent in the first half of the 1980s, and about 3 per cent in the second half of the 1980s (Table 3.1). This loss of money to the rural sector was probably more than compensated for by the sharp increase in farm-gate prices, a vast expansion of the rural credit system, and the investments made by RTVEs (Table 3.6). But dispersed into the hands of peasant households farming small, scattered plots of land, surplus income was quietly spent on housing and consumer goods, without any money being invested in non-profitable grain production (except direct subsidies by the RTVEs), but was to find its way instead into lucrative rural industries, ranging from processing plants for agricultural products to simple machine shops.

Lack of state investment was also important in explaining the onset of stagnation of grain production between 1985 and 1988, but more significant was the failure of price reforms to offer a consistent incentive to the producers. Although the procurement price of grain gradually increased throughout the 1980s, the rate of growth suddenly slowed down after 1985 and the input prices increased dramatically (these were induced by the policy changes of 1985). Therefore, the difference between the free market and procurement prices widened again after 1985. Table 4.2 shows survey results of the local market and procurement prices and their difference in Bolo county. The free market price of grain increased at an annual rate of 28 per cent between 1985 and 1987 but the mixed procurement price increased at a much lower rate. As a result, the difference between these two prices increased gradually from 3.87 yuan/100 *jin* in 1984 to 15.84 yuan/100 *jin* in 1987.

The rapid increase in the market price of grain may have been due to increasing costs of modern inputs. Table 4.3 shows the composition of production costs of grain and their changes for Bolo county. The unit product cost increased rapidly. The annual growth rate of unit production cost increased from 1.91 per cent between 1965 and 1978 to 21.1 per cent in 1985–86. After 1978, the costs of all inputs increased considerably, especially the costs of materials. The price of fertilizers, for example, increased by 134 per cent from 1978 to 1986. Sudden

Table 4.2 Grain procurement and market prices in Bolo county (yuan/50kg)

	1984	1985	1986	1987
(1) free market	20.00	24.00	30.00	35.00
(2) mixed procured	15.06	16.13	16.13	19.16
(1)–(2) = difference	4.84	7.87	13.87	15.84
(1)/(2) = ratio	1.30	1.50	1.90	1.80

Source: Wei (1987), Table 3.

Table 4.3   Changes of grain production costs in Bolo county

	Costs (yuan/50 kg)				Growth rate (%, p.a.)		
	1965	*1978*	*1985*	*1986*	*1965–78*	*1978–85*	*1985–86*
Total	8.60	11.01	13.23	16.03	1.91	2.60	21.10
Labour	5.00	5.32	5.85	6.43	1.94	5.10	5.60
Materials	3.60	5.70	7.37	9.59	5.23	7.60	25.10
Fertilizer	1.38	1.47	4.28	5.17	1.98	2.80	16.20

Note: The cost of fertilizer is included in the cost of materials.
Source: Wei (1987), Table 4.

increases in input prices were largely due to the abandonment of input subsidies which had been maintained until 1985. The policy changes in 1985 were meant to introduce *more price liberty*, resulting in reduced production profitability as the increase in grain procurement price was not kept in line with the increase in input prices.

The increasing gap between the procurement and free market prices of grain made grain production much less profitable as opposed to the production of non-grain crops and sideline activities. Although changes in the agricultural tax system and reduction of the procurement quota since 1985 had enabled farmers to divert resources (mainly land) from grain production to non-grain crops with relatively high profitability, the contraction of grain area not only reduced domestic grain supply but also undermined production efficiency due to misallocation of scarce resources.

As pointed out at the beginning of this chapter, grain self-sufficiency is one of the most important goals of state agricultural policy. If the government still wants to emphasize grain self-sufficiency and is unwilling to spend foreign exchange on grain imports, some new methods have to be introduced.

There are two ways to restore the incentive of grain producers or force them to produce more grain. One is to abolish the *contracted* procurement and *indirect* tax on grain, letting farmers sell their grain to the market. This may damage the urban consumers if they cannot receive any food subsidy from the state and have to purchase food from the free market. If the state still tries to keep the consumer price at the original level by purchasing grain from the free market at the market price and selling it to the consumers at the original price, state budgetary costs will inevitably expand. Expanding budgetary costs caused many difficulties for the government in the late 1970s and early 1980s. The question here is whether the government is willing to enter the same trap created in the earlier stage of reforms. The changes made in 1985 were partly aimed at cutting the budgetary cost at the expense of farmers. It is doubtful that the government will follow the same path of large budgetary cost in the future as it did in the 1978–84 period.

The other alternative is to let the consumers share some of the costs of price reforms by cutting grain subsidies. Improving the diets of urban consumers in the last decade actually reduced the consumption of foodgrain. Abundant evidence has shown that the supply of state-subsidized grain to urban consumers surpassed their actual needs. People in the urban areas have been seen selling grain coupons on the streets for eggs and vegetables.

According to a survey conducted by an undergraduate student, Lin Hao Chien, of the South China Agricultural University, residents in the Western district of Guangzhou, capital of Guangdong Province, purchased from the state food stores 25.9 per cent less than they were entitled to at the subsidized price in 1981. The difference was 23.63 per cent in 1982, 17.96 in 1983, 14.34 per cent in 1984. A national survey conducted by the Rural Development Research Centre of the state council in 1989 shows consistent results. The sample covers 1579 households randomly selected from a number of cities in Liangnin, Shanxi, Guangdong and Hubei provinces. The results suggest that almost half of the households buy grains from the local free markets. In 1989, the average amount of grain bought from the state grain stores was only 9kg per month per capita, 4.5kg or 33 per cent lower than the state rationing of 13.5kg (Research Centre of Economic Policy, MOA, 1991, p. 196).

In theory, reducing subsidized supplies to the urban dwellers is not only possible but also desirable. Firstly, the current state rationing is substantially higher than what is actually needed. Secondly, the current share of expenditure on grain as a proportion of total household expenditure is very small. Even if state rationing is totally abolished, an average urban household can afford to buy food at the current market prices. According to the same survey by the Rural Development Research Centre of the State Council, although 58 per cent of total household expenditure was on food in 1989, only 11 per cent of total expenditure was for grain.

However, grain subsidy has been a politically sensitive issue and it requires time and consistent effort for careful planning to avoid social and political unrest. This is why the government has been reluctant to make significant changes in the urban subsidy programme in order to reduce state budgetary costs and increase grain output and farm incomes. An economic reason is that food subsidy can help keep urban wages low and maintain the competitiveness of industries – a complicated but obvious mechanism to transfer resources from agriculture for industrial development.

Supposing the state does not want to liberalize the grain market and change the grain subsidy programme, there is one more alternative which can be used to improve grain output. This will, unfortunately have to be similar to that practised before 1978: compulsory procurement with a significant *indirect* tax. With compulsory procurement at the same low prices, grain output may increase, at least in the short run. However, the consequence will be similar to those in the pre-reform period. Local self-sufficiency of grain and less specialization of agricultural production will be inevitable. Production efficiency and farmers' incomes will deteriorate.

In short, whatever alternative is used, it is always a problem of balancing the interests of the urban consumers, the farmers and the state budget. There is also a problem of cost and benefit to the society of each policy package. Therefore, any alternative policy must depend on the objectives of the government and its willingness to accept the consequences or the social costs of policy changes.

The policy changes made in the late 1980s provide a vivid example of how government has made its choice to stimulate grain production after four years of recession. The changes are characterized by two points: (a) moderately increasing the procurement prices of grain and promising instant payments to farmers; and (b) re-emphasizing the role of compulsory quotas. The results have been satisfactory so far as grain output is concerned. After four year of stagnation, grain output started to rise significantly in 1989 and reached two new records in 1990 and 1991 (Chapter 2, Table 2.11). However, the changes introduced are not based on long-term planning: they can at best be regarded as tentative methods to boost production. At worst, they will create more difficulty in sustaining the long-term growth of grain output in the future. The paradoxes are similar to the changes made at the beginning of economic reforms. Increasing procurement prices is desirable and effective for the farmers but it will increase the financial burden on the state. In the long term, sustainable price reform should be accompanied by the gradual reduction of consumer subsidies. Reinforcement of the mandatory quota is useful to raise production in the short run, but will recreate negative effects similar to those in the pre-reform era, such as increased inefficiency of resource allocation, and lower farmers' incomes. A compulsory quota is effective and beneficial because it ensures higher production and reduces efficiency losses in a situation where the procurement price is lower than the free market price. The mechanism and rationality of the interaction of these two policy instruments (low procurement price and compulsory quota) is the focus of quantitative exercise in this book. The reduction of the production quota in 1985 at a time when the gap between free market and state prices was increasing was particularly inimical to grain production and allocative efficiency. The policy change in that aspect can be regarded as putting the cart before the horse.[5]

4.2.2 Constraints on future growth

In China, the area sown to grain has suffered a significant decline in absolute terms and as a share of total crop area (see Table 4.4, Figure 4.1). Thus output growth has been predominantly determined by the level of yield. It is obvious that major constraints to grain production are land resources and the ability to raise yield.

A recent report of the World Bank (1991) identifies four major constraints to grain production in China: *physical constraints*; *technical constraints*; *institutional constraints* and *economic constraints*. The following discussion follows a similar line of argument but stresses the importance of relative profitability between crops and the effect of non-farm enterprises in the rural areas.

(1)   Resource and physical constraints

These are the most important constraints on agricultural and grain production in China since independence. Recent reforms have imposed a much greater downward pressure on the area sown to grain crops.

Although China has a huge territory of more than 9.6 million square kilometres, only 10.4 per cent of it can be used for agricultural production (Table 4.5).

Increasing land constraint is reflected by a steady decline in arable land and increase in population since independence. Table 4.6 and Figure 4.2 reveal the acute shortage of per capita arable land and crop area over time.

Table 4.4   Crop areas in China, 1949–91

Year	Crop area (mil. hectares)				As % of total area		
	Total	Grains	Industrial crops	Others	Grain	Industrial crops	Others
1952	141.26	123.98	12.49	4.78	87.77	8.84	3.39
1953	144.04	126.64	11.66	5.74	87.92	8.09	3.99
1954	147.93	128.99	12.42	6.51	87.20	8.40	4.40
1955	151.08	129.84	13.99	7.26	85.94	9.26	4.80
1956	159.17	136.27	14.70	8.13	85.61	9.23	5.11
1957	157.24	133.63	8.46	9.15	84.98	9.20	5.82
1958	151.99	127.61	13.77	10.62	83.96	9.06	6.98
1959	142.40	116.02	13.55	12.83	81.47	9.51	9.01
1960	150.58	122.43	12.91	15.23	81.31	8.58	10.12
1961	143.21	121.44	9.32	12.45	84.80	6.51	8.69
1962	140.23	121.62	8.76	9.85	86.73	6.25	7.02
1963	140.22	120.74	10.19	9.28	86.11	7.27	6.62
1964	143.53	122.10	11.99	9.44	85.07	8.35	6.58
1965	143.29	119.63	12.21	11.45	83.49	8.52	7.99
1966	146.76	120.99	12.24	13.60	82.44	8.34	9.27
1967	144.94	119.23	12.20	13.51	82.26	8.42	9.32
1968	139.83	116.16	11.35	12.32	83.07	8.12	8.81
1969	140.94	117.60	11.43	11.91	83.44	8.11	8.45
1970	143.49	119.27	11.71	12.47	83.12	8.16	8.69
1971	145.68	120.85	11.93	12.9!	82.95	8.19	8.86
1972	147.92	121.21	12.53	14.18	81.94	8.47	9.58
1973	148.55	121.16	12.80	14.60	81.56	8.61	9.83
1974	148.64	120.98	12.89	14.77	81.39	8.67	9.94
1975	149.55	121.06	13.40	15.08	80.95	8.96	10.09
1976	149.72	120.74	13.72	15.26	80.64	9.17	10.19
1977	149.33	120.40	13.53	15.40	80.63	9.06	10.31
1978	150.10	120.59	14.44	15.08	80.34	9.62	10.04
1979	148.48	119.26	14.77	14.45	80.32	9.95	9.73
1980	146.38	117.23	15.92	13.22	80.09	10.88	9.03
1981	145.16	114.96	17.56	12.64	79.20	12.10	8.71
1982	144.75	113.46	18.79	12.50	78.38	12.98	8.63
1983	143.99	114.05	17.76	12.19	79.20	12.33	8.46
1984	144.22	112.88	19.29	12.05	78.27	13.37	8.35
1985	143.63	108.85	22.38	12.40	75.78	15.58	8.64
1986	144.20	110.93	20.29	12.99	76.93	14.07	9.01
1987	144.96	111.27	20.72	12.96	76.76	14.30	8.94
1988	144.87	110.12	21.50	13.25	76.02	14.84	9.15
1989	146.55	112.20	20.99	13.36	76.56	14.32	9.12
1990	148.36	113.47	21.42	13.48	76.48	14.44	9.09
1991	149.59	112.31	23.47	13.80	75.08	15.69	9.23
Annual Growth Rate(%):							
1952–57	2.17	1.51	−7.50	13.85	−0.64	−9.46	1.11
1958–62	−2.26	−1.87	0.71	1.47	0.41	3.04	1.04
1963–78	0.43	−0.05	3.17	2.70	−0.48	2.73	1.02
1952–78	0.23	−0.11	0.56	4.51	−0.34	0.32	1.04
1979–89	−0.22	−0.65	3.46	−1.09	−0.44	3.68	0.99
1952–89	0.10	−0.27	1.41	2.81	−0.37	1.31	1.03

Sources: ZGNCJJTJDQ (1949–1986), 1989; ZDNYTJZL (1987–91), various issues.

Figure 4.1   Structural changes in crop areas, 1952–91

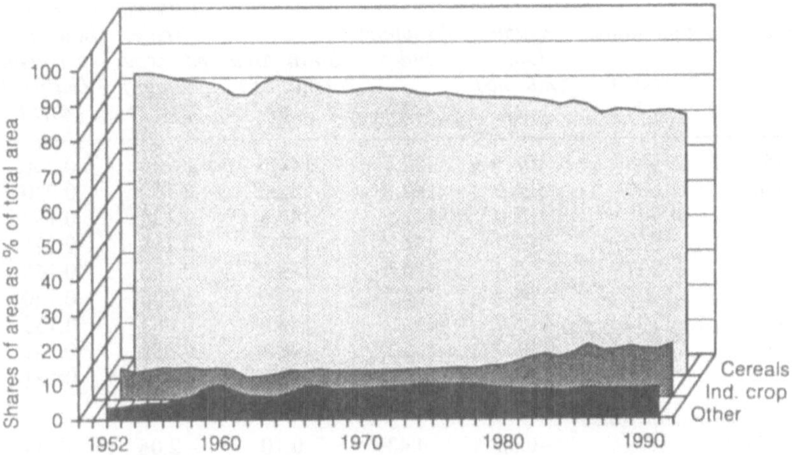

Source: Table 4.4.

Table 4.5   A summary of land utilization

		% of total land area
1.	Cultivated land	10.4
2.	Fruit, tea and other perennial crops	0.3
3.	Forestry	12.7
4.	Grassland	33.0
5.	Urban, roads and industrial areas	6.9
6.	Inland waterways and shallow seas	3.1
7.	Stone, bare ground, deserts, marshes, etc.	19.4

Source: Chen and Buckwell (1991), Table 3.1.

Due to land erosion and urban development, total arable land decreased by 11.8 per cent from 108.5 to 95.7 million hectares between 1953 and 1989, with a negative growth rate of 0.29 per cent per year. Although the total area under crops remained almost constant over time, this was possible because of an increasing cropping index. However, as population grows, per capita arable land and crop area decline rapidly. In 1953, per capita arable land was 0.185 hectares. By 1989, it was only 0.087 hectares, with a rate of decrease of over 2 per cent per year.[6] Per capita crop area declined from 0.245 to 0.133 hectares over the same period, with a negative growth rate of 1.69 per cent per year.

Table 4.6   Land–population balance

Year	Population (mil.) (1)	Arable Land (mil. ha) (2)	Cropping Index (3)	Sown Area (mil. ha) (4)	Arable per head (ha/p.c.) (2)/(1)	Sown area per head (ha/p.c.) (4)/(1)
1953	588	108.5	132.7	144.0	0.185	0.245
1958	660	106.9	142.2	152.0	0.162	0.230
1966	745	103.0	142.5	146.8	0.138	0.197
1971	852	100.7	144.7	145.7	0.118	0.171
1976	937	99.4	150.6	149.7	0.106	0.160
1983	1025	98.4	146.3	144.0	0.096	0.140
1989	1104	95.7	153.2	146.6	0.087	0.133
1990	1123	95.7	155.1	148.4	0.085	0.132
1991	1142	95.7	156.3	149.6	0.084	0.131
Annual Growth Rate						
1953/91	1.76	−0.33	0.43	0.10	−2.06	−1.63

Sources: Guo Shutian (1991), Table 4, p. 29; Tables 2.8 and 4.4; *ZGNYTJZL* (1991), p. 2, p. 12 and p. 30.

Figure 4.2   Per capita arable land and crop area (ha/pc)

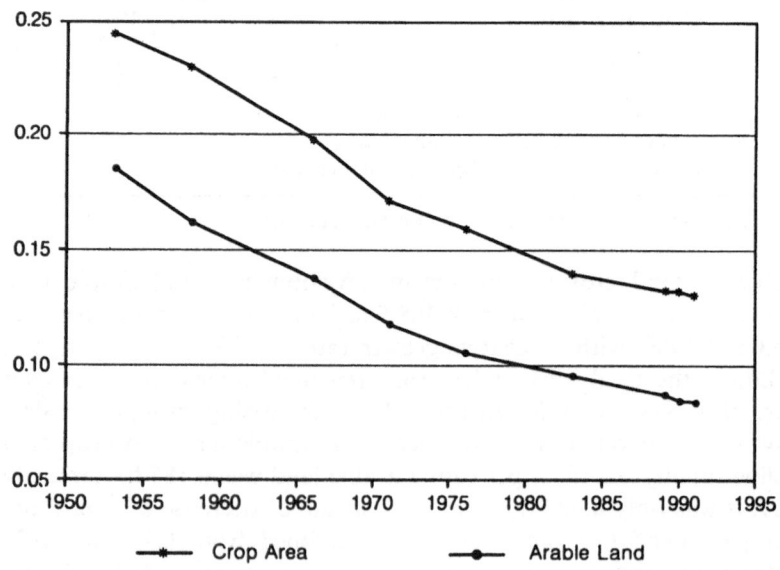

Source: Table 4.6.

The main physical constraints are variable rainfall and the frequent occurrence of droughts in the north, the heavy rains accompanying typhoons which lead to flooding and waterlogging in lower lying and poorly drained areas, and the cold weather in the north and high altitude areas that restrict the growing seasons. Current reports indicate that parts of the irrigation and drainage system, vital to increasing yields and stabilizing production, have deteriorated. This is mainly due to inadequate state investment in rural infrastructure. Without significant improvement in the physical and environmental conditions in vulnerable areas, the land productivity of many marginal areas will remain low.

(2) Technical constraints

In general, China has developed a good technical base to increase and maintain yields and production. For most grains, yields are already high by international standards (Table 4.7). Input levels and yields are generally much higher in China than in other poor developing countries in Asia. For instance, in 1989, grain yield and the amount of fertilizer used per hectare in China were more than twice as high as in India. The levels of inputs, such as fertilizers and irrigation, and grain yield in China are even higher than the corresponding average levels of all the developed countries, though they are still considerably lower than those in the most productive countries such as Japan and some of the western European nations.

This indicates that additional increments in grain yields may be difficult to achieve without much higher marginal costs. The main constraints seem to be lack of sufficient disease and water-logging resistance in wheat and lack of sufficient disease and pest resistance in rice varieties. This may be restricting yield response to the higher rates of fertilizers, especially nitrogen fertilizers, now being applied and may be a main cause of yield stagnation in wheat and rice. Another important constraint appears to be lack of information to plan future fertilizer requirements and to increase the efficiency of the use of organic manure and chemical fertilizer. The use of manure fertilizer has been rapidly replaced by the use of chemical fertilizer, primarily due to the increasing opportunity cost of labour in collecting manures and other organic fertilizer (see Chapter 2, the village experience, and Fan, 1991). This shift of fertilizer use can be very damaging to soil fertility in the long term.

Table 4.7 Comparisons of cereal outputs and inputs between China and the rest of the world

Countries	Population growth (%, pa) 1971–80	Population growth (%, pa) 1980–89	Cereal output per cap (kg) (1989)	Fertilizer use (kg/ha) (1988)	Tractor/ arable land (unit/k.ha) (1988)	Irrigation ratio (%) (1988)	Cereal yield (kg/ha) (1989)
World	1.8	1.8	358.3	98.7	18.8	16.7	2646
Asia	2.0	1.9	269.4	114.8	12.1	33.9	2686
China	1.7	1.3	327.1	147.8	9.4	46.4	4014
Japan	1.2	0.6	116.3	415.1	475.9	69.3	5671
Indonesia	2.3	1.8	280.6	112.8	0.9	47.5	3757
India	2.1	2.2	230.2	65.2	4.5	25.2	1851
Africa	2.8	3.0	144.5	22.5	3.3	6.6	1228
Egypt	2.4	2.7	194.7	400.0	19.1	100.0	5524
Europe	0.5	0.3	585.5	229.3	81.5	13.7	4246
UK	0.1	0.2	392.3	345.7	74.8	2.2	5771
France	0.6	0.5	1018.3	311.6	83.1	7.5	6068
N.C. America	1.6	1.4	844.0	83.7	21.3	9.7	3627
Canada	—	1.0	1827.0	47.0	16.5	1.8	2186
USA	1.1	1.0	1145.8	93.6	24.9	9.6	4410
S America	2.5	2.2	267.4	41.1	9.7	7.6	2073
Brazil	2.7	2.2	296.5	47.5	10.8	3.9	1995
USSR	0.9	0.9	704.6	117.0	11.8	9.1	1905
N AMERICA	—	1.4	1211.0	84.5	23.2	8.1	3846
W EUROPE	—	0.3	512.6	225.4	104.0	14.0	4503
OCEANIA	—	1.2	1128.3	38.5	8.7	4.5	1665
DEV.PING M E	—	2.5	213.2	50.4	6.3	19.0	1844
AFRICA	—	3.1	122.3	10.8	2.0	3.8	1053
LAT AMERICA	—	2.2	233.0	49.0	9.3	10.5	2067
CENTR PLANND	—	1.2	431.4	166.7	14.3	19.8	2980
E EUR + USSR	—	0.8	739.4	136.7	16.2	9.8	2268
DEV.PED ALL	—	0.7	707.5	124.6	32.2	9.8	3089
DEV.PING ALL	—	2.1	248.7	76.8	6.4	20.6	2346

Notes and Source to Table **4.7**
Notes: DEV.PING M E = middle-income developing countries.
       CENTR PLANND = centrally planned economies.
       DEV.PED ALL = all developed countries.
       DEV.PING ALL = all developing countries.
       pa = per annum; per cap = per capita; ha = hectare; k.ha = 1000 hectares;
       Irrigation ratio = percentage of arable land under irrigation.
Source: FAO (1990).

(3)   Institutional constraints

Institutional constraint is a serious impediment to achieving higher growth in yield and production. Recent reports have indicated that inadequate and untimely supplies of fertilizers and agrochemicals impose a significant constraint on grain production. This is serious in a production system heavily dependent on intensive modern inputs and high standards of farm management to achieve high yield. Poor timing of application due to late arrival of inputs reduces effectiveness and potential return on investments. Insufficient extension services and inadequate supplies of new seed varieties impose additional constraints on production. It is reported that the government seed corporation can only meet 20 per cent of total seed requirement. Most government resources are focused on hybrid rice and corn production. Lack of investment, pricing policy and inadequate operating margins to maintain viable seed corporations are the main problems. The agro-technical extension bureaux (responsible for extensive services) face serious difficulties in bringing new technology to the majority of the agricultural households. Estimates suggest that only 20 to 50 per cent of known technology is being applied in the 2400 counties and 70,000 townships that the service is supposed to cover (World Bank, 1991, p. 16). Only 14,000 of the 70,000 townships have proper facilities and trained personnel. It has been argued that there is a lack of viable farmer extension groups, lack of good extension methodology and lack of adequate operating budget. In addition, the incentives for the qualified scientists graduating from different universities to work in the rural areas have been traditionally low due to harsh working conditions and lack of financial rewards. China has produced tens of thousands of agricultural scientists every year since 1981, but only a small proportion of them are working in the rural areas.

Farm mechanization poses problems among smallholders due to small farm size, the large number of small plots, the lack of suitable agricultural machinery, and the lack of spare parts and fuel. Currently 91 per cent of tractors are owned by farm households and 9 per cent (mostly medium and large tractors) by collectives. There are 37,000 mechanization stations operated by collectives or collective group farmers who can provide service and repair facilities to individual households. However, it appears that inadequate provision has been made in the system to provide for depreciation and replacement costs. It is estimated that 1.2 million small tractors and 160,000 medium and large tractors need replacing (World Bank, 1991). A reduction in tractor power could have a marked effect on cropping intensity, which is particularly important for a system in which absolute arable land cannot be increased. The lack of tractor power for agricultural purposes may be exaggerated by the diversion of tractor use for non-farm transportation and other business. Even in the most developed areas in China, the level of mechanization in agriculture is still very low.

(4)   Economic constraints

Economic constraints are mainly induced by policy biases against agriculture, and grain production in particular. Reduced state investment in agriculture during the whole reform period is undoubtedly a major constraint for future agricultural growth and improvements in grain yields. Many official reports published or issued in the late 1980s indicate that rural infrastructural and irrigation conditions have been deteriorating, as is reflected by the reduced ability of the farming sector to cope with natural disasters such as droughts and flooding. This is a disappointing outcome for a period in which agriculture and grain production have experienced the highest growth rates after independence.

This can be blamed for the lack of concerted reform methods implemented during economic reforms. At the local government level, there are often strong incentives to use revenue to invest in enterprises offering a higher return than agriculture. At the farm level, small farm size and low farm-gate prices (for cotton and cereal crops) make it difficult to generate sufficient income to adopt the high input technology needed to achieve high yields. In addition, partial market reforms have allowed farmers to sell non-grain products at free market prices while the prices of cereal and cotton are still under state control.

State price control on cereals, in general, discourages incentives to grain producers because even the negotiated prices currently offered to farmers are generally lower than the local free market prices.[7] To secure a certain amount of procurement, quantity control is still effectively in place. This therefore imposes a negative effect on regional specialization, leading to increased inefficiency of resource allocation.

The second negative effect of state pricing on grains arises from its nature in being pan-territorial, or partially pan-territorial. This is why in a good harvest, farmers in the low-cost provinces (such as Anhui, Henan, Jiangxi and Shangdong) would find it difficult to sell their products to the state at the official price, while farmers in other parts of the country would still find it unprofitable to sell any of their product to the state at the same price. Pan-territorial pricing, under this situation, is inimical to all farmers in all provinces. If state pricing is more flexible so that the costs of transportation between deficit and surplus regions are taken into account, the crop pattern can be shifted towards achieving greater spatial efficiency. As a result, farmers in surplus provinces would have no difficulty in selling their products although the price may be lower. Those in deficit regions would not be obliged to deliver their grains at an obvious loss.

The shift in crop pattern is a dynamic process. It may result in a short-term financial loss to producers in the low-cost areas. However, the negative effect of lower prices may be offset by a more secure marketing outlet. Historical evidence indicates that the main constraints on grain production in surplus regions have been the incapacity of the state to procure the surplus at a predetermined price because of inadequate storage facilities, lack of funds, and inadequate transportation to ship grain quickly from the surplus to the deficit areas, rather than prices. Thus a more effective way to stimulate grain production in the low-cost regions (in most cases they are also the less developed areas) is through public investments on transportation and storage to increase inter-regional marketing efficiency.

Lack of storage facility is demonstrated by the fact that, on average, about 20–35 million tons of grain have to be stored in the open air every year. In addition, the state grain marketing authority has to rent a storage capacity equivalent to several million tons from other systems. To house such an amount of grain, the state needs an additional storage capacity of 30–50 million tons. Furthermore, most existing warehouses are very old and in poor condition. Of the total storage capacity currently in use, an equivalent of 55 million tons constructed during the 1950s and 1960s needs to be replaced (Research Centre of Economic Policy, MOA, 1991, p. 175).

Another constraint on state procurement is the low profitability of the grain marketing firms. Before 1988, the profits earned by these firms through selling grains and oils at negotiated prices were equally divided between the state and the firms. In 1989, the marketing firms were allowed to retain only 40 per cent of their profits. By 1990, the profit share to the firms was reduced to 20 per cent. This low profitability of grain marketing firms has also been exacerbated by higher interest on bank credit. Before 1988, grain marketing firms could borrow at lower interest rates: with subsidized rates of 3.96 per cent per year, and negotiated rates of 7.92 per cent. In 1988, credit subsidies to these firms were abolished. From September 1988, marketing firms had to pay the free market rate of 9.0 per cent. In February 1989, the interest rate was further increased to 11.34 per cent (Research Centre of Economic Policy, MOA, 1991, p. 174). Higher interests on credits greatly discouraged their incentives to buy grains from the farmers.

The marketing situation since (e.g. 1989–91) has been very similar to that during 1983–84.[8] It gives an impression that China may have *too much grain*. In fact, the country is still importing in large quantity. This *structure surplus* is created purely by the inadequacy of the marketing system. It is very likely that grain production will go into stagnation again if no appropriate measures are taken to solve the marketing problems.

A longer-term solution to prevent the vicious cycle occurring in grain production (stagnation—dramatic short-term incentives—high output—difficulty in marketing—dramatic short-term disincentives—stagnation), is to increase the ability and flexibility of state pricing and marketing policy. If state pricing and procurement policy is sufficiently flexible, the grain marketing situation can be significantly improved without extra budgetary outlays. The short-cut solution to this depends on the flexibility of pricing, both seasonally and spatially. A second mechanism is frequent adjustments of state procurement quotas imposed in different provinces. The long-term solution has to rely on a systematic reform in both the consumer and producer markets, and improvements in the transportation and storage conditions.

4.3 PROSPECTS OF GRAIN PRODUCTION

4.3.1 Arguments for food self-sufficiency

Given the limited land resources and the need to produce other crops, one may argue that it may be more efficient to substitute more imports for domestic production. Most officials and economists in China have taken a different view. Food self-sufficiency is still regarded as the first priority in the current agricultural policy. According to the Ministry of Agriculture (and the central government), food self-sufficiency is justified by the following arguments.

(a) A rapid reduction of world food reserves in the last few years.

During 1987–8, total world food reserves decreased by 30 per cent from the previous year. In 1988–89, global food output increased by 3.4 per cent, but the reserves declined by 5 per cent. Total reserves were only 17 per cent of the global demand, which was the lower limit of food security reserves (17–18 per cent). It is expected that the supply situation in future will be more pessimistic. If China starts to import in large quantities, it will exercise a significant effect on the world market and raise international food prices considerably.

(b) It is not efficient for China to import grain in large quantity as grains produced domestically are much cheaper than those imported from abroad.

In 1989, the c.i.f. prices of wheat from the United States averaged $170 per ton. The free market price of wheat in the domestic markets was equivalent to $150 per ton. In addition, foreign exchange is still a major constraint to development. Even if imported grains were as cheap as those produced domestically, it would be more desirable to encourage domestic production.

A recent World Bank report also indicates that domestic producers, particularly in the surplus areas, have been severely *taxed*. The Effective Protection Coefficients (EPCs) are substantially lower than one for all grain crops if the shadow exchange rates and free market prices are used to calculate the coefficients (Table 4.8). The results suggest that domestic producer prices can be significantly increased without losing China's competitiveness of food production in the international market. Table 4.8 also indicates that rice is taxed more severely than the other grains.

China's comparative advantage in grain production can also be measured by another indicator, the domestic resource cost (DRC). The same World Bank report estimates the DRC coefficients for rice, wheat, corn and soybeans, calculated at the official exchange rate and the shadow exchange rate in 1988 (3.71 and 6.50 yuan per US$ respectively). The national average DRCs are presented in Table 4.9. In general, China has a considerable advantage in grain production, especially rice production.

(c) From the long-term point of view, large imports of grain will induce stagnation of grain production and aggravation of rural unemployment.

Table 4.8   EPCs for grains (1988)

	Official exchange rate (3.71¥=$1)		Shadow exchange rates (6.5¥=$1)	
Commodities	EPCa	EPCb	EPCa	EPCb
Milled Rice				
at unit export price	0.40	0.90	0.23	0.51
Wheat				
at unit import price	0.96	1.50	0.53	0.85
Corn				
at unit import price	0.74	0.98	0.42	0.56
at unit export price	0.83	1.08	0.47	0.61
Soybeans				
at unit import price	0.63	—	0.35	—
at unit export price	0.59	—	0.33	—

Notes:
(1)   The Effective Protection Coefficients (EPCs) are calculated as the ratio of the respective domestic price net of fertilizer and equipment production costs at the domestic prices to the border price net of fertilizer and equipment production costs at the border prices. A more correct calculation should use the parity prices rather than the border prices. Thus the values presented in this table have underestimated the EPCs if import prices are used, and overestimated if export prices are used.
(2)   EPCa are calculated by using state contract prices. EPCb are calculated by using the free market prices in the surplus producing areas.
(3)   The unit export price is used for rice because rice is an exporting good. The unit import price is used for wheat because wheat is an importing good. Both export and import prices are used for corn and soybeans because these two products can either be importable or exportable.
Source: World Bank (1991), Working Paper no. 7, vol II, Table 19, p. 31.

(d)   From the security point of view, international food prices generally suffer from significant fluctuations and the supply may not be reliable. International trade in grains has been regarded as a politically sensitive issue. China's policy makers will find it highly undesirable to rely on imports to make up a significant proportion of domestic demand. In addition, grain imports to China have mainly been from the United States which has heavily subsidized grain exports. If, for political or economic reasons, export subsidies are abolished, grain imports will suddenly become too expensive.

Table 4.9   DRCs for major food crops in 1988

	Rice	Wheat	Maize	Soybeans
At official rate (¥3.71 = $1)				
Total production cost (y/ton)	269	340	229	430
Unit import price (y/ton)	—	441	409	906
Unit export price (y/ton)	671	—	372	955
Cost of imported inputs (y/ton)	37	72	42	42
DRC at import price	—	0.72	0.51	0.45
DRC at export price	0.37	—	0.57	0.42
At shadow rate (¥6.5 = $1)				
Total production cost (y/ton)	269	340	229	430
Unit import price (y/ton)	—	774	716	1587
Unit export price (y/ton)	1176	—	653	1673
Cost of imported inputs (y/ton)	59	108	68	64
DRC at import price	—	0.35	0.25	0.24
DRC at export price	0.19	—	0.28	0.23

Notes: The DRCs, which approximate the cost in domestic resources of earning a net dollar in foreign exchange (where the latter deducts the cost of imported inputs), is defined in this table as the ratio of (a) the total cost, at domestic prices, of producing one ton of the commodity, net of the cost of imported inputs valued at border prices, to (b) the border price of the commodity (i.e., unit import or export prices), net of the cost of imported inputs valued at border prices. Fertilizer and equipment use are considered to be imported inputs in the calculations.
Source: World Bank (1991), Working Paper no. 7, vol. II, Table 20, p. 32.

4.3.2 Potential for future growth

Significant fluctuations of grain production during the 1980s have two distinctive signals: the vulnerability of the production and marketing system and the great potential of grain output. The first signal is discouraging in the sense that inappropriate state pricing and marketing policies can easily lead to a vicious cycle of stagnation, recovery and new record output. The second signal is encouraging in the sense that it is possible for China to maintain a high level of food self-sufficiency.

There are a number of factors which can contribute to maintaining and increasing grain production. According to the Ministry of

Agriculture, resource constraint is not absolutely insurmountable, and production potential still exists in the following aspects.

(a) There are 13 million hectares of land which can be reclaimed for agricultural production. If 75 per cent of the land can be effectively opened up, the total cultivated area can be increased by 9.75 million hectares, or about 10 per cent.

(b) Two-thirds of the current cultivated area of 95.3 million hectares are classified as medium and low productivity areas. If the average productivity of all these areas could be improved up to the current national average output of 3.15 tons per hectare, the total national output could be increased up to 600 million tons.

     This can be achieved through investments (irrigation and other infrastructure), improving farm management and more intensive use of modern inputs and technologies. It is reported that only 30 to 40 per cent of current farming technologies are used. Although the input levels are rather high compared to other countries, the use of modern inputs in the poor areas is still low. In addition, there is a great potential to increase land productivity (even in the high yield regions) through the adoption of new technologies and farming techniques.

(c) The current multiple-cropping index is about 156 per cent. It is estimated that the maximum potential is 160 per cent. If the cropping index is increased by 1 per cent, the crop area will be increased by almost 1 million hectares.

4.4 SUMMARY

Grain production and agricultural growth have not been smooth throughout the reform period. The period of 1979–84 was characterized by high growth in production and farmers' incomes. The period from 1985 to 1988 was characterized by the stagnation of grain production and a slow growth in agriculture and farmers' incomes. Grain and other agricultural production improved during 1989–91, despite the devastating and widespread flooding in 1991.

     Fluctuations in grain production and agricultural growth during economic reforms are the outcome of inconsistency in policy changes, especially the changes made in 1985. The fundamental problem lies in the dilemma faced by the government when trying to balance the interests of farmers, urban consumers and the budgetary cost. Large increases in producer prices of agricultural commodities shortly after reforms increased the budgetary costs considerably as consumer subsidized prices remained unchanged. The policy changes in 1985 were intended to cut the budgetary costs by: (1) reducing input subsidies to farmers; (2) slowing down the increase of procurement prices; and (3) cutting mandatory quotas. All these factors had direct negative effects on grain production.

The recovery of grain production, and therefore agricultural growth, during 1989–91, was due to two policy changes: (1) more price incentives, e.g. the procurement price of grain was raised by 18 per cent in 1989; and (2) reinforcement of mandatory quotas. These two changes were not accompanied by reform in the consumer market. Thus budgetary cost increased again, and it was reflected by the difficulty of farmers in the surplus regions in selling grains to the state – a situation similar to that in 1983 and 1984.

If there are no appropriate measures taken to solve the marketing problems, it is likely that grain production will slump into another stagnation as in 1985. The incapacity of the state marketing system has been characterized by: (1) lack of storage facilities; (2) low profitability of the marketing agencies; and (3) inadequacy of transportation.

All the above problems have been caused by the need to transfer resources from agriculture for urban and industrial development and the inability of government to effectively adjust the speed and level of such transfers. After more than forty years of industrialization, agriculture is still sacrificed to support state industries and urban development. This is the most disappointing aspect of the economic development and industrialization in China since independence. It will take much more time before farmers can be entirely freed from exploitation. Any short- and medium-term policies have to take this into account. For the foreseeable future, the problem will be to decide what resources can be transferred out of agriculture, and at what rate. The contribution of agriculture to economic development will therefore be the focus of discussion in Chapter 5.

Part II
Theoretical Analyses of Agricultural Policies

Introduction

This part of the book presents some theoretical analyses on the role of agriculture in economic development, government policy objectives and major policy instruments. As many agricultural policies and policy objectives in China are very similar to those in other developing countries, the analyses in this part of the book are not exclusive to China, but are also relevant for most other developing countries.

Chapter 5 discusses the role of Chinese agriculture in economic development since 1949. Based on the classification by Kuznets (1964), and Ghatak and Ingersent (1984), four contributions of agriculture to national economic development are identified: product contribution, market contribution, factor contribution and foreign exchange contribution. In the early stage of development, as agriculture was the predominant sector of the economy, these four contributions were highly significant. Due to some inevitable laws of economic development, the role of agriculture in economic growth declined over time in terms of its contribution to GDP, employment, market and foreign exchange earnings. However, the development of non-agricultural sectors was greatly dependent on resource transfers from agriculture.

Chapter 5 also discusses why unbalanced development between the agricultural and non-agricultural sectors is inevitable and why it is desirable to transfer resources from agriculture to support industrial development. However, we also argue that the speed of resource transfer from agriculture must be *appropriate* so that both agriculture and industry can grow concomitantly. Excessive resource transfers from agriculture not only damage agricultural growth in the long run, but also encourage wasteful investment in the non-agricultural sectors. The success of industrial development does not solely depend on the ability of government to squeeze agriculture for capital. It also depends on many other factors, including industrial efficiency, the purchasing power in agriculture, the ability of agriculture to provide raw materials for industrial production, and the ability of agriculture to generate foreign exchange.

A comparison between the pre-reform and reform periods supports our argument. In the pre-reform period, agriculture was excessively exploited by the government to support the ambitious industrialization programme. Industrial output grew much faster than agriculture.

Rapid growth in industry, however, did not help to improve people's living standards and employment. This was mainly due to wasteful industrial investments and production inefficiency. In the reform period, the effort to readjust the economic structure has helped to stimulate faster agricultural growth. At the same time, industrial growth has been also higher than in the pre-reform period. Although superior performance in all the economic sectors during the reforms can be explained by institutional changes and market reforms, there is no doubt that structural adjustments in the economy to emphasize the role of agriculture and slow down the rate of resource transfer from agriculture is one important reason for the success of reforms since 1978.

Economic growth in China during the reform was by no means smooth. There have been a number of fluctuations since the reform. In some years, e.g., 1978–79, 1984–85 and 1987–88, the economy (mainly the industrial sector) expanded at a rate which was not sustainable. In other years, such as 1980–81, and 1989–90, radical methods had to be adopted to slow down growth deliberately. The vicious cycle of development in the Chinese economy, characterized by the pattern of *stable growth–unrealistic expansion–radical adjustment–stable growth*, has been repeated at least three times since 1978. The reasons for the frequent reappearance of this cycle can be attributed to the inability of the government to manage the economy. However, one major cause of the problem lies in the fact that policy-makers tend to be short-sighted with regard to resource allocation between different sectors of the economy. Due to the nature of agricultural production, investments in this sector cannot generate quick results as in industry and other non-agricultural sectors. Thus, rapid economic growth in the short term is always accompanied by excessive exploitation of agriculture and large but often wasteful investments in industry. The long-term effect of this is naturally undesirable.

The problem of uneven development over time can be explained by the lack of understanding of the inconsistencies of various agricultural contributions, the lack of a set of coherent policy objectives, and the lack of understanding of the complicated relationships among these conflicting objectives.

The inconsistencies of various agricultural contributions are analysed in detail at the end of Chapter 5. Policy objectives and their relationships are presented in Chapter 6.

In China, the objectives of agricultural policy are directly or indirectly related to the national industrialization programme. Among these objectives, food self-sufficiency, price stabilization involving significant consumer subsidies, and export earnings are the most important ones. To realize these objectives, a set of policy instruments are required. The objective of these instruments is how they can be used effectively to tax agriculture and subsidize the rest of the economy.

In practice, a number of instruments can be used to achieve the same objective. For instance, to tax agriculture, the Chinese government has employed a set of related instruments, including direct taxation, indirect taxation through controlling domestic prices and foreign exchange rate, import/export taxation or subsidy. However, different policy instruments may have different effects on production, consumption and economic efficiency. Thus, the choice of the right instruments is critical for the government. The Chinese government, like many other governments of the developing countries, have also employed a set of quantitative methods as secondary instruments to reduce the negative effects of price instruments. For instance, rationing is used as a secondary instrument to reduce the social cost of consumer food subsidy, and mandatory production quotas are employed to reduce the social cost of indirect production taxation.

To support the theoretical discussion, Chapter 7 presents some empirical results using mathematical models. For presentation purposes, the principle and formulation of mathematical models are presented in the self-contained Appendix in Part III. The discussion in Chapter 7 summarizes the simulation outcomes of the mathematical models. In addition, it discusses the limitations of our quantitative approach and makes proposals for future improvement. It evaluates the policy implications of the computational results and suggests future grain policies in China.

5 Agriculture's Role in the Chinese Economy

5.1 INTRODUCTION

Government agricultural policies are often related to the 'expected' contribution of agriculture to the national economy. Throughout the post-independence period, agriculture has been treated as the primary source of capital accumulation to build up a capital-intensive and modern industrial sector. To understand fully the frequent shifts in agricultural policies, it is important to analyse the contribution of agriculture to the development of the national economy.

In the first part of the book some dilemmas of agricultural policies, especially those concerning grain production, were discussed. One of these was concerned with the adjustment of indirect taxation on grain producers. In the pre-reform period, because procurement prices of grain and other crops were far too low, farmers were subject to severe taxation. In the reform period, due to dramatic increases in prices, the government incurred increasing budgetary costs as consumer food subsidies were kept unchanged. An attempt to reduce the budgetary cost would involve increasing taxes on farmers if urban consumers and the state-industrial sector are to continue to be protected.

The problems of agricultural development and grain production are related to the fundamental deficiency in the national economy. That deficiency lies in the many insurmountable problems of the state-owned industrial sector. After more than forty years of development and protection, the industrial sector still requires heavy subsidies from agriculture.

In this chapter, it is argued that 'appropriate taxation' on agriculture is desirable and essential for industrial development. However, it is also argued that 'excessive taxation' is inimical not only to grain production and the rural economy, but also to the process of industrialization. This is because the contribution from agriculture is not just through resource transfers, but also includes production, market and export. Excessive resource transfers will inevitably reduce the capacity to produce, the ability to buy industrial goods, and the potential to export of the agricultural sector.

Many authors have discussed the role of agriculture in economic development. According to Kuznets (1964), it can be defined in terms of three contributions, i.e., product contribution, market contribution and factor contribution. Ghatak and Ingersent (1984) follow the same definition but they add foreign exchange as another contribution by agriculture as agriculture is a net foreign exchange contributor in most of the poor nations.

The discussion in this chapter follows the same classification made by Ghatak and Ingersent and described below.

(1) **Product contribution:** Expansion of the non-agricultural sector is strongly dependent on domestic agriculture, not only for a sustained increase in the supply of food, but also for raw materials used in manufacturing products such as textiles.

(2) **Market contribution:** Because of the strong agrarian character of the economy during the early stages of economic growth, the agricultural population inevitably forms a substantial proportion of the home market for producers' goods as well as consumer goods.

Market contribution is not as obvious as product contribution but its importance cannot be understated as 80 per cent of all consumers in China are farmers. Their ability to buy will determine the pace of development of the domestic industries, especially for those producing products which are not tradable or competitive in the international markets.

(3) **Factor contribution:** Because the relative importance of agriculture in the economy inevitably declines with economic development, agriculture is seen as a principal source of capital for investment elsewhere in the economy. Thus the development process involves the transfer of surplus capital from agriculture to the non-agricultural sector. Similarly, development also entails the transfer of surplus labour from agriculture to non-agricultural occupations, especially in the long term.

In China factor contribution of agriculture has been dominated by capital transfer but not so much by the provision of cheap labour from agriculture to industry (or from rural to urban), owing to the deficiency of the industrial structure. However, labour transfer from traditional agriculture to non-farm enterprises within the rural economy since economic reforms has been very significant.

(4) **Foreign Exchange Contribution**:   Domestic agriculture is capable of contributing to the balance of payments either by augmenting the country's export earnings or by expanding the production of agricultural import substitutes (Ghatak and Ingersent, 1984, Chapter 3).

   In China, the importance of agricultural export is reflected in its very high share (directly and indirectly) in total exports and the large net foreign exchange earnings.

We now consider the four contributions one by one.

5.2   PRODUCT CONTRIBUTION

As seen from Table 2.1, the industrial and agricultural sectors contribute 74–87 per cent of the total national income in China. However, the share of industrial output over time has been increasing whereas that of agricultural output has been decreasing.

   For example, the share of industrial output in total national income increased rapidly from 19.5 per cent in 1952 to 48 per cent in 1989. The share of agricultural output declined from 57.7 per cent to about 32 per cent over the same period (Table 2.1).

   Growth of the agricultural sector can be expected to lag behind industrial growth in a developing economy like China, where per capita incomes are rising. There are a number of reasons which explain this phenomenon. Some of these reflect an inevitable trend of economic development, such as Engel's Law and the increasing dependence of agriculture on industries for inputs. Slow growth in agricultural output can also be explained by government policies which discriminate against agriculture.

   Despite a decreasing share of agricultural output in the national economy, agriculture's product contribution is still critical in sustaining economic growth for two reasons. Firstly, the entire population still relies on agriculture for food and many industries rely on agriculture for material inputs. Secondly, agriculture is a large market for the industries producing modern agricultural inputs and consumer goods.

5.2.1 Why is agriculture's role diminishing?

There are three possible reasons: (1) inelastic demand for foods and agricultural products as incomes grow; (2) increasing dependence of agricultural growth on industrial inputs; and (3) government policies.

(1) The demand for food and other agricultural products is generally less income-elastic than the demand for industrial goods and services (Engel's Law).

Engel's Law states that as incomes increase, consumers tend to spend a smaller proportion of incremental expenditures on food and agricultural products. Development experience of the industrial nations shows that an increasing proportion of consumers' expenditure is spent on luxury industrial goods and services. The disparities in living standards between different countries are often measured by the expenditure on food as a proportion of total household expenditure. In general, if people spend most of their income on food, they are considered to be 'subsistence' consumers.

In China, the term 'subsistence' has been interpreted as 'having enough clothing and food'. The rate of progress away from subsistence depends on the growth of per capita income. In the pre-reform period, Chinese farmers were trapped in subsistence because there was no significant per capita income growth. By 1978, almost 68 per cent of household expenditure was spent on food, 13 per cent on clothing and only a very small share on housing and other non-necessities. As income grows, the share of household expenditure on food and clothing declines rapidly. This was shown during the reforms. More income was spent on housing and other luxury goods. By 1989, the share of expenditure on food had declined to only 54 per cent, or by about 14 per cent over eleven years. The share of expenditure on clothing declined to about 8 per cent, or by more than 4 per cent over the same period (Table 2.5).

Incomes in the urban sector also increased during the reforms and the share of expenditure on food and clothing declined over time as well (Table 2.4).

Both food and clothing are directly or indirectly produced by the agricultural sector. As the share of a consumer's yuan spent on these goods declines over time, the consumption linkage to agriculture will diminish, and a gradual shift of development effort from agriculture to non-agricultural sectors becomes desirable. However, the fact that food consumption still accounts for more than 50 per cent of consumers' expenditure in both the rural and urban areas suggests that agriculture is still an important engine of growth.

It has to be pointed out that Engel's Law is much more obvious and significant in the case of basic food items, especially cereals. Although the share of food expenditure has been declining in recent years in China, the share of expenditure on some particular items such as meat, fish, cooking oil and some other processed food items may have increased. On the other hand, the share of expenditure on cereals may have decreased much more rapidly. Available statistics do not provide enough information on individual food items but a recent World Bank study indicates that demand for grains has been declining, particularly in the urban sector.

Urban direct consumption of grain declined by 6 per cent from 145kg per capita in 1981 to 137kg per capita in 1988. Direct grain (processed) consumption was 167kg per capita in 1957, 22 per cent higher than in 1988. In the rural areas, direct grain (unprocessed) consumption remained almost constant at about 260kg per capita between 1980 and 1988, increasing from 257kg in 1980 to 267 in 1984 and then declining to 260 in 1988. The weighted average of the urban and rural direct grain consumption indicates that national average direct per capita consumption of grain increased by 26 per cent from 193kg in 1970 to 257 in 1984 and then declined by 3 per cent to 249 in 1988 (World Bank, 1991, p. 6).

The changes in direct grain consumption in China suggest that: (1) urban people have been above 'subsistence' since the 1950s; (2) rural people were below 'subsistence' prior to economic reforms but have reached or exceeded subsistence levels since economic reforms.

Engel's Law pertaining to grain consumption can be measured by the values of income elasticities of grain consumption. If the income elasticities are lower than one, the Law is verified. Table 5.1 summarizes the estimates in the World Bank (1991) study on grain demand in China cited above. Although the estimates are not subjected to rigorous statistical tests, income elasticities in the table provide a rough indication which is consistent with the above conclusions.

In both the time-series and cross-sectional analyses, income elasticities of direct grain consumption in the urban sector are significantly negative (suggesting that grain is an inferior good). In the time-series analysis, income elasticity of direct grain consumption in the rural sector is also negative but it is positive in the cross-sectional analysis.

Table 5.1 Income elasticities of direct grain consumption (physical consumption)

	Urban	Rural
Time-series (1981–88)	−0.20	−0.06
Cross-sectional (1988)	−0.33	0.15

Source: World Bank (1991), pp. 9–12.

Decline in direct grain consumption per capita has been more than offset by increasing indirect grain consumption. Indirect consumption of grain as animal feed is estimated to have tripled from 28kg per capita in 1970 to 95kg per capita in 1988. In sharp contrast to the expected continued decline in direct per capita grain consumption, expected strong demand for animal products will continue to push up indirect demand for feed grain.

Strong indirect consumption of grain is reflected in the very high and positive income elasticities of demand for various animal products for both the rural and urban population (Table 5.2).

Table 5.2 Income elasticities of demand for animal products, 1988

	Urban		Rural	
	Time series	Cross section	Time-series	Cross-section
Pork	0.19	0.54	N.A.	N.A.
Beef and mutton	1.50	0.72	N.A.	N.A.
Pork/beef/mutton	0.34	0.57	0.73	0.26
Poultry	1.66	0.93	1.48	2.16
Eggs	0.52	0.58	1.06	1.66
Fish	0.04	0.59	0.89	4.54

Note: Significant differences between the time-series and cross-sectional results suggest that the estimates are very crude. They have to be interpreted with great caution.
Sources: Data: SSB (1989) household income and expenditure data.
Estimates: World Bank (1991), p. 12.

Increased indirect demand for grain in China is not only reflected in increased demand for animal feed, but also in increased demand for grain used by the industrial sector to produce beverages and alcoholic drinks. Thus, total per capita grain demand is likely to increase in the future, albeit with a constant or decreasing per capita direct consumption of grain.

Notwithstanding the positive and much higher income elasticities of demand for meat and fish, Table 5.2 indicates that all non-cereal products, except poultry, are income inelastic in the urban areas. In the rural areas, the demand for 'all red meat' (pork/beef/mutton) is also income inelastic, but for poultry products and fish it is elastic.

International comparisons also suggest that grain and most other agricultural products are income inelastic: in other words Engel's Law is verified. Table 5.3 presents the income elasticities of demand for selected agricultural commodities in several countries.

Another point which has not been discussed so far is the change of income elasticities over time. It is almost certain that demand for cereals and other agricultural products will become less income elastic in countries where real per capita income grows. In countries where real per capita income is stagnant or decreasing, demand for foods will remain more income elastic.

Table 5.3   Income elasticities of demand for agricultural products in selected countries

	Egypt 1974/5		India 1973/4		Indonesia 1978		Mexico 1977	UK 1980	USA 1972/3
	Urban	*Rural*	*Urban*	*Rural*	*Urban*	*Rural*			
Cereals	0.15	0.61	0.21	0.48	0.15	0.23	−0.16	0.01	0.08
Sugar	0.75	1.26	0.66	1.33	N.A.	N.A.	0.02	−0.26	0.16
Vegetables	0.52	1.33	0.83	0.78	0.67	0.66	0.42	0.12	−0.05
Fruit	0.94	2.09	1.39	1.48	1.66	0.86	1.21	0.49	N.A.
Meat	0.98	1.74			0.98	1.29	1.02	0.10	0.34
Fish	0.76	1.50	0.8	1.14	1.17	1.01	0.99	−0.01	N.A
Eggs	1.07	2.01			1.08	1.92	0.57	−0.09	−0.54
Milk	N.A.	N.A.	1.06	1.59	0.57	0.06	0.18		
Fats/oils	0.84	1.48	0.70	0.99	N.A.	N.A.	0.17	−0.03	−0.32
Total food	0.75	1.28	0.79	0.82	0.74	0.72	0.09	—	—

Source: Colman and Young (1990), Table 6.2.

Low and decreasing income elasticities of demand for grain and other agricultural products have a number of implications for development planning. The empirical estimates above confirm that income elasticities for food decline as income grows and the proportion of total expenditure on food will decrease. Hence the focus of economic activity will gradually shift away from the agricultural sector. An increasing degree of urbanization would have the same effect. Furthermore, the empirical results for individual food products imply that, as income grows, the pattern of production within the agricultural sector will have to change. As expenditure on some products (namely livestock products, fruit and vegetables) increases more rapidly than on staple foods (grains), increased specialization in livestock and horticultural production will be required. This has been verified by the structural change within the agricultural sector in China during the last decade. The share of crop output in total agricultural output has declined whereas that of livestock output has increased rapidly since 1978 (Chapter 2, Table 2.13).

(2) Due to land constraints, any increase in agricultural productivity becomes increasingly reliant on modern technology, especially on modern inputs such as fertilizers and machinery purchased from the non-agricultural sector of the economy.

Agricultural growth in China throughout the whole post-independence period has been achieved under increasing pressure on arable agricultural land (Chapter 4, Table 4.6 and Figure 4.1). Agricultural growth has increasingly depended on increased use of industrial materials (for a detailed discussion, see the market contribution of agriculture, Section 5.3). As a result, the share of agricultural output becomes less significant.

(3) Government intervention can also affect agricultural growth

Apart from the above two factors which can result in an inevitable decline in the importance of agriculture in economic development, government intervention can also lead to slower agricultural growth. This was a distinct phenomenon throughout the pre-reform period in China.

According to the above arguments, a gap between agricultural and industrial growth rates is probably inevitable and desirable provided it is not artificially distorted by government policies. However, the Chinese experience suggests that if the gap is too large, the whole economy will suffer.

The enormous gap between the growth rates of agriculture and industry in China in the pre-reform period (Chapter 2) cannot be wholly explained by the above two arguments. In fact, demand for agricultural products in the pre-reform period must have been very elastic as people's incomes were low. The failure to meet this demand through encouraging faster agricultural growth is one major reason for poor economic performance.

Drawing on historical evidence, many Chinese economists have pointed out that agricultural growth, especially grain output, has a direct and strong impact on the overall performance of the economy (Zheng Zhong, 1988; Guo Shutian, 1991). Zheng Zhong concludes that throughout modern Chinese history since 1949, there existed a positive and significant linkage between the growth rates of agriculture and the whole national economy (Zheng Zhong, 1988).

This conclusion is confirmed in the discussion presented in Chapters 2 and 3. All the information and analysis point to the importance of balancing the growth rates of different sectors. An inevitably diminishing share of agricultural output in total GNP does not mean that industries can develop on their own. The success of economic reforms has narrowed the gap between the growth rates of agriculture and industry.

However, even during economic reforms, development has been by no means smooth. The tendency on the part of the government to undermine agriculture still remains. In the early years of reforms (1979–84), food output and agricultural production increased rapidly.[1] However, as discussed in Chapter 3, grain output stagnated for four years between 1985 and 1988) and agricultural growth did not meet planners' expectations. In retrospect, the stagnation of food production and unsatisfactory performance of the agricultural sector as a whole during 1985–88 can also be attributed to the continuation of government's discriminatory policies against agriculture. Agricultural investment has been reduced in order to finance large industrial projects throughout the reform period. The benefits of increased prices have been gradually eroded since 1985 by the increasing costs of current inputs such as fertilizers and other chemicals. The profitability of food production will continue to decline if the increase in food procurement prices cannot be kept in line with the increase of input costs.

5.2.2 Why is agriculture still important?

Although the agricultural sector will become relatively less important in the long run, it is essential not to overlook the critical importance of product contribution of domestic agriculture to the maintenance of an adequate rate of economic growth in the short term. Development planners and policy-makers in China have time and time again opted for a strategy of rapid industrialization without paying sufficient attention to agricultural growth.

There are two basic reasons why the development of the national economy is constrained by the rate of growth in agricultural output. Firstly, the domestic farm sector is an important source of raw materials for use in industries such as textiles and food processing, as well as being the principal source of food for consumption by growing numbers of the non-food producers employed in industry. In China, the gross output of agriculture-based industries increased by more than 12 times between 1952 and 1985, accounting for more than two-thirds of the light industrial sector and about one-third of total national industrial output (Table 5.4). As agriculture-based industries are generally more labour intensive, their contribution to employment is undoubtedly far more important.

Table 5.4 Output values of agriculture-based industries and their shares in total industrial output values

Year	Gross output values of agriculture-based industries (bil. yuan)	As % of gross light industrial output values	As % of gross industrial output values
1952	19.4	87.5	56.4
1960	40.7	74.0	24.6
1965	50.4	71.7	36.1
1970	73.6	70.0	32.6
1975	97.6	70.1	30.4
1980	160.4	68.5	34.2
1985	275.7	67.0	33.2

Notes: Gross output values are measured at periodical constant prices: before 1957 at 1952 prices, before 1970 at 1957 prices, before 1980 at 1970 prices and before 1986 at 1980 prices.
Source: ZGNCJJTJDQ (1949–86), pp. 60–1.

Secondly, as agriculture becomes more and more integrated with other sectors of the economy, the multiplier effects of increased agricultural production and income assume an increasing importance in relation to the growth in demand for the products of domestic industry and the associated demand for labour and other industrial inputs (see Section 5.3).

The most important product contribution of Chinese agriculture to the national economy is food. Diversification of the economy is contingent on domestic food producers producing a surplus in excess of their own subsistence, which is large enough to feed a growing number of non-food producers.

Although, in principle, shortfalls in domestic food supplies can be made good by increasing food imports, in practice such imports are severely restricted by the scarcity and high cost of foreign exchange. As noted above, expenditure on food for an ordinary household consists of more than half of its total expenditure. Shortage of food due to stagnant domestic food production and/or inability to expand imports will lead to higher prices. Due to the primacy of food as a wage good, rapid food price inflation frequently leads to serious social and political instability which is inimical to economic growth. Because higher food prices are likely to result eventually in higher industrial wages, industrial product prices can also be expected to rise after a time-lag. With food and industrial product prices both higher, a rise in the general price level or inflation is virtually inevitable. Thus a process of cost-push inflation may be visualized in which lagging supplies and rising prices of food provide the initial impetus, but other prices follow after a time-lag.

Although there are many reasons for the unprecedentedly high inflation in China in the late 1980s (1985–89), the shortfall of domestic grain production between 1985 and 1988 may be considered to be the initial impetus for the general inflation. The efforts in 1989 to contain inflation featured strong measures to increase grain output. Subsequent bumper harvests in 1990 and 1991 greatly helped to reduce the inflationary pressure experienced in the previous years. By the end of 1991, inflation dropped to the lowest level in almost a decade, thanks to abundant supplies of foodgrains in all rural markets.[2]

5.3 MARKET CONTRIBUTION

Market contribution includes the so-called backward linkage and consumption linkage of agriculture to the rest of the economy. The backward linkage refers to the demand by agriculture for different inputs, such as fertilizers, chemical insecticides, machinery, electricity, transportation, and rural infrastructure. The consumption linkage refers to the multiplier effect of farmers' consumption on the rest of the economy. The purchasing power of farmers is contingent on the pace of agricultural development.

5.3.1 Backward linkages

In China, agricultural demand for industrial inputs has been accelerated by the increasing constraints of arable land and crop area. As discussed in Chapter 4, per capita arable land and crop area declined respectively by 2.08 per cent and 1.69 per cent per year between 1953 and 1989.

Despite the acute shortage of land, China has managed to achieve significant growth of per capita agricultural output, especially during the reform period. Per capita grain output increased by 0.44 per cent during 1953–78 and by 1.56 per cent during 1978–89 per year (Table 2.11). Per capita agricultural output value increased by 0.7 per cent per annum during 1953–78 and 4.67 per cent during the period 1978–89 (Chapter 2, Tables 2.10 and 2.11).

Sustainable growth has been achieved through steady improvements in land productivity. For example, grain yield increased by 250 per cent from 1028kg/hectare in 1949 to 3608kg/hectare in 1984.

Increased land productivity, however, has relied on the increasing use of industrial products, such as fertilizers, insecticides, farm machinery and equipment, electricity and many others. This suggests that an increasing share of agricultural output value is accounted for by the non-farm sectors.

Recent econometric analyses by Fan (1991) and Yao and Xie (1991) indicate that agriculture and grain production have relied more and more on chemical fertilizers, machinery and irrigation. The elasticities of these inputs with respect to agricultural output or grain yield are highly significant and increasing over time (Table 5.5; also see Chapter 3, Table 3.5).

The use of chemical fertilizers

The use of chemical fertilizers in China can be classified into three rather different periods of development.

Table 5.5 Production elasticities for selected inputs

Year	Fan[a] (agricultural output)		Yao and Xie[b] (grain yield)		
	Chemical fertilizer	Machinery	Chemical fertilizer	Irrigation	Machinery
1965	0.140	0.078	0.114	0.196	0.039
1970	0.181	0.127	0.170	0.175	0.038
1975	0.221	0.176	0.211	0.185	0.060
1980	0.262	0.225	0.438	0.165	0.064
1985	0.303	0.274	0.491	0.131	0.043

Notes:
[a] The dependent variable is total national agricultural output value. Fertilizer is total national fertilizer input. Machinery is total national machinery input value.
[b] The dependent variable is grain yield per hectare of crop area. Fertilizer is physical input of fertilizer (effective contents) per hectare of crop area. 'Irrigation' is measured as the ratio of irrigated area to total crop area. 'Machinery' is the ratio of area under tractor ploughing to total crop area. Detailed data and definitions are in Table 5.6 of this chapter.
Sources: Fan (1991); Table 2; Yao and Xie (1991), Table 6 (and Table 5.6).

The first period, between 1949 and 1964, was characterized by a low level but high growth rate of use. The level of application was only 0.15 kg per hectare in 1949. With an annual growth rate of 31 per cent, it increased up to 9kg per hectare. During this period, grain production was mainly sustained by intensive use of organic fertilizers.

The second period of development was characterized by medium levels of application and by a still high but lower rate of growth. This period covered the Cultural Revolution (1965–78). The level of fertilizer application increased from 13.5 to 59.3kg per hectare, with an annual growth rate of 11 per cent. During this period, the use of organic fertilizers was still important but its dominant position was gradually supplemented by chemical fertilizers.

The third period of development started from the late 1970s. It is characterized by high levels of application and lower growth rates. The level of application increased from 66.8kg per hectare in 1979 to more than 160kg per hectare in 1989, with an annual growth rate of about 8 per cent (Table 5.6).

The development of irrigation

Compared with other modern inputs, irrigation in China has a long history. In fact, a well-developed irrigation system is a precondition for sustainable growth of grain and agricultural production. As early as the 1930s, about 26.5 million hectares, or 27 per cent of China's total arable land, were under effective irrigation. Irrigation suffered a serious decline during the war in the 1940s, so that by 1952, with some recovery from the war, only 18 per cent of arable land was irrigated. Since then, the area under irrigation has expanded rapidly. Many irrigation and drainage projects were carried out during the Great Leap Forward and the Cultural Revolution. By 1970, 36 million hectares, or 40 per cent of cultivated land, were being irrigated.

An important development in modern China is the increasing share of irrigation powered by machinery and electricity. In the past, most irrigation pumps were operated manually. The rapid expansion of tube wells and electric irrigation pumps signifies the scale of investment. Between 1952 and 1974, the power-irrigated area in China rose from 0.32 million hectares to 32 million hectares, an increase of 100 times. The share of power-irrigated area in total irrigated area rose from 1.6 per cent to 52 per cent over the same period. Powered irrigation is more effective and labour-saving, but it depends on the increasing inputs provided by the industrial sectors.

The development of agricultural mechanization

Agriculture was highly unmechanized before independence. Significant development of agricultural mechanization started in the Great Leap Forward period. By 1965, the total machine ploughed area had reached 15.6 million hectares, or six times greater than in 1957. Total machinery power increased from 1.65 million horsepower in 1952 to 15 million in 1965. The number of large and medium-sized tractors increased by almost 55 times within 13 years, and that of harvesting machines increased from 284 to 6704 units.

Further development of agricultural mechanization was pursued during the 1970s. By 1980, there were 745 million large and medium-sized tractors and 1.67 million small-sized tractors (less than 15 horsepower per unit). There were also 27,045 combine harvesters and 70,000 trucks in the rural areas. The machine-ploughed area reached 41 million hectares, or 41 per cent of total cultivated land (Table 5.6).

The pace of agricultural mechanization in China has been greatly constrained by the pressure of rural underemployment. Recent

Table 5.6   Fertilizer use, irrigation and mechanization indices

Year	Grain yield (kg/ha)	Fertilizer application (kg/ha)[a]	Irrigated area/ Arable land	Tractor plough area/ Arable land
1949	1035	0.45	0.08	0.00
1950	1155	0.75	0.09	0.00
1953	1320	0.75	0.20	0.02
1955	1425	1.65	0.22	0.02
1956	1410	2.03	0.22	0.02
1960	1170	4.50	0.27	0.02
1964	1575	9.00	0.31	0.15
1965	1635	13.50	0.32	0.15
1970	2010	24.75	0.35	0.18
1971	2070	24.75	0.36	0.21
1972	1980	28.50	0.38	0.22
1973	2190	34.50	0.39	0.27
1974	2280	33.00	0.41	0.29
1975	2355	36.00	0.43	0.33
1976	2370	39.00	0.45	0.35
1977	2355	42.00	0.45	0.39
1978	2535	59.25	0.45	0.41
1979	2790	66.75	0.45	0.42
1980	2745	87.00	0.45	0.41
1981	2835	92.25	0.45	0.37
1982	3135	104.25	0.45	0.36
1983	3405	115.50	0.45	0.34
1984	3615	120.75	0.45	0.36
1985	3480	123.60	0.45	0.36
1986	3525	134.25	0.46	0.38
1987	3630	137.90	0.46	0.40
1988	3635	147.80	0.46	0.43
1989	3690	160.86	0.47	0.45

Note: [a] Fertilizer use is calculated as effective contents.
Sources: ZGNCJJTJDQ (1949–1986), Department of Planning, Ministry of Agriculture, P. R. C. Beijing Agricultural Press, Beijing, May 1989; ZGNYTJZL (1987, 1988, 1989), Department of Planning, Ministry of Agriculture, PRC, Beijing Agricultural Press, Beijing; Yao and Xie (1991), Table 3.

economic reforms have helped create many non-farm employment opportunities and have pushed up rural wages. It is likely that further improvement in farm productivity will depend more and more on machinery inputs.

In short, agricultural growth is not only dependent on increased use of fertilizers, irrigation and machinery but also on many other non-farm inputs, such as improved seeds, insecticides, services and transportation. In China, the share of material inputs in total agricultural total output value increased from 21 per cent in 1957 to 32 per cent in 1986 (*ZGNCJJTJDQ*, 1949–86, p. 126). It indicates that any given rate of agricultural growth will generate a higher rate of growth in the industries producing these materials.

5.3.2 Consumption effects

Farmers' income is largely determined by the rate of agricultural growth. As farmers' income grows, they will be more able to buy consumer goods which are produced not just by the agricultural sector itself but also by the non-agricultural sectors, such as housing, clothing and electronic products.

This indirect effect has been widely recognized. Mellor (1976), Mellor and Lele (1973) and Hazell and Roell (1983), in particular, have called attention to the potential power of agricultural consumption linkages. They conclude that middle-sized peasant farmers – to a much greater extent than their large-scale and urban counterparts – spend incremental income on labour-intensive, rurally-produced goods, thereby generating important second-round demand multipliers. Hirschman's (1958) early indictment of agriculture as a low-linkage, underpowered engine of growth erred, according to Mellor (1976), because it ignored these important agricultural consumption linkages.

As China is still basically agrarian and 80 per cent of the total population is living in the countryside, the rural community is inevitably a dominant market for many domestically produced consumer goods.

Economic reforms have brought about sustainable and significant increase in farmers' incomes. As rural incomes grow, farmers are spending a higher proportion of incremental expenditure on manufacturing products. For instance, many traditional durable goods, such as TV sets, watches and bicycles, have a limited market in the urban areas, but they have a very strong and expanding market in the countryside (see Chapter 2 for a discussion of change in expenditure structure for the rural households).

5.4 FACTOR CONTRIBUTION

There are two basic factors, i.e., capital and labour, which can be provided by agriculture to the national economy. Whereas agriculture's product contribution derives from agricultural production per se, and the market contribution derives from trade with other sectors, the factor contribution derives from resource (capital and labour) transfers to industry and other non-agricultural sectors of the economy.

5.4.1 Capital contribution

From the experiences of many developed countries such as the United Kingdom, Japan and the former Soviet Union, it can be seen that agriculture has been the most important capital contributor in the early stages of industrialization. In many developing countries, including China, capital formation has been largely dependent on agriculture. Before discussing alternative ways of transferring capital from agriculture to other sectors, and criteria used to judge the appropriate amount or rate of transfer, let us explain briefly why the net transfer of capital from agriculture is a credible means of economic development.

There are four arguments in favour of transferring capital out of agriculture.

(1) Even assuming that capital:output ratios in agriculture and non-agriculture are identical, the incremental demand for capital in the non-agricultural sector may be higher in a developing economy because the demand for non-agricultural products and services is generally more income-elastic than the demand for food and other agricultural products (see Section 5.2).

(2) In a developing economy like China, the non-agricultural sectors are much more capital-intensive than the agricultural sector although scope exists for raising productivity in agriculture by using more capital.

(3) Agriculture is virtually the sole domestic source of savings and investment during the initial stages of development.

(4) Farmers are likely to benefit indirectly from non-agricultural-type investments (e.g., improvements in communication and trans-

port). With the exception of an extreme case described by a dualistic economy with a wholly uncommercialized agriculture, indirect benefits should also derive from growth in the industrial sector. For instance, if industrialization can raise average living standards, it should also benefit farmers.

As discussed in Chapter 1, in China there are two alternative means of capital transfer from agriculture to the rest of the economy. One is the direct agricultural taxation on land which is termed the 'direct' transfer of resources. This kind of capital transfer has been decreasing in China over the last several decades.[3]

The other is the 'indirect' transfer of resources, which may include price controls on agricultural products, agricultural inputs and consumer goods, indirect taxes and exchange rate manipulation. All these methods share the common objective of changing the inter-sectoral terms of trade (i.e., the farm:non-farm product price ratio) against agriculture. In China, the discrimination in the terms of trade against agriculture is called the 'scissors difference'.[4]

As the importance of direct agricultural taxation in resource transfer has been diminishing (Chapter 1), indirect taxation, or the 'scissors difference', is a dominant form of capital transfer from agriculture to the rest of the economy – mainly the state-owned industrial sector.

After the socialist reforms in 1956, all the former private industrial enterprises were virtually transformed into public enterprises owned by the state. In the 1950s, the economy was more or less similar to that described by the conventional dualistic models (Lewis, 1954; Fei and Ranis, 1964). The industrial sector was very small while agriculture played a dominant role in the economy. Due to the influence of the former USSR, the Chinese government tried to build up a heavy industrial sector beginning with the Great Leap Forward. The development of heavy industry required a good deal of investment which, in turn, required a high accumulation rate. To maintain high accumulation, the government had to opt for a low-wage policy in the state industrial sector (Chapter 1). In fact, wages paid to workers were so low that they could not maintain a subsistence level of living without some other forms of payment. State subsidies, such as housing benefits, food subsidies and free medical care, therefore became necessary for state employees.

The food subsidy programme covered all the non-agricultural population (basically the urban dwellers). As the non-agricultural population expanded, the budgetary cost increased rapidly. To reduce the pressure on the state budget, the government decided to let the farmers share the costs of food subsidies from the late 1950s. The means to impose this was to introduce compulsory procurement and pay very low prices to farmers for foodgrains and many other agricultural goods.

The costs of food subsidies had been largely borne by the farmers until 1979 when the new agricultural policies increased the procurement prices of all agricultural products.

The degree of discrimination against agriculture, as measured by the 'scissors difference', was modest in the early 1950s. The price of agricultural goods was only about 10 per cent lower than their real market values and those of industrial goods higher by the same magnitude (Table 5.7). However, discrimination against agriculture was greatly intensified from 1957 onwards. The 'scissors difference' increased up to 25.8 per cent in 1957 and reached its peak of 27 per cent in 1978, suggesting that for any amount of trade between agriculture and the rest of the economy, farmers had lost more than one-quarter of the exchanged values. Economic reforms have significantly improved agriculture's terms of trade but, by 1983, agricultural goods were still paid for at 24.6 per cent lower than their real market values, and the 'scissors difference' was 18.5 per cent (Table 5.7).

Table 5.7   The 'scissors difference' in selected years

	1952	*1957*	*1978*	*1983*
(1) The prices of industrial goods higher than their actual market values by %	9.7	14.8	21.2	14.8
(2) The prices of agricultural goods lower than their actual market values by %	9.6	20.4	37.1	24.6
(3) The 'scissors difference' by %[a]	9.7	25.8	27.0	18.5

Note:
[a] The 'scissors difference' measures the extent to which agriculture's terms of trade are discriminated against for those goods which are exchanged between agriculture and the rest of the economy (also see footnote 4).
Sources: Guo Shutian (1991), p. 64.

In the late 1950s and early 1960s, more than two-thirds of the state budgetary incomes were from indirect transfer of capital from agriculture. Although, in relative terms, indirect transfer of capital declined to less than 50 per cent of total state budgetary incomes, in absolute terms the amount of transfer increased dramatically in the 1980s (Table 5.8).

Although part of state budgetary revenue flows back to the rural areas in the form of agricultural and rural infrastructural investments, input subsidies, rural social services (education and health care), etc., total state subsidies to the rural areas are much smaller than that extracted from the farmers. Thus, even by deducting the inflows into the rural sector, the resulting net transfer from agriculture through indirect taxation was about 20 per cent of the total state budgetary revenues between 1978 and 1985 (the net transfer must have been more severe in the pre-reform era).

Capital transfers in Table 5.8 do not include taxes on rural township enterprises. Taxes collected from the township enterprises are substantial; they increased from 2.2 billion yuan in 1978 to 36.5 billion

Table 5.8   Capital transfers from agriculture (billion yuan)[a]

	Total transfers[b]	As % of state budget	State subsidy to agriculture[c]	Net transfer
1957	22.1	71	—	—
1965	35.8	76	—	—
1971	38.4	52	—	—
1978	44.2	39	—	—
1984	67.7	—	—	—
1986	89.4	—	—	—
1987	104.5	—	—	—
1988	130.2	—	—	—
1953–85	680.0	—	—	—
1978–85	481.5	47.5	224.9	176.8

Note:
[a] Values are calculated at current prices.
[b] Calculated according to the 'scissors difference' and the volumes of trade between agricultural and non-agricultural sectors.
[c] State subsidies to agriculture include state investment in agriculture, input subsidies and all the other possible subsidies to agriculture and the rural areas.
Source: Economic and Policy Research Centre, MOA (1991), no. 2, pp. 139–41, 159.

yuan in 1989, or from 4.2 per cent of the total state tax revenues in 1978 to 13.4 per cent in 1989 (Table 3.6). If these were included, rural-to-urban capital transfers would be much greater in the 1980s.

Increasing procurement prices of food and other agricultural products has been one major component of the new agricultural policy since 1979. In principle, the increase in procurement prices is equivalent to reducing the indirect transfer of capital from agriculture to the rest of the economy.

The reduction of indirect agricultural taxation, however, had become a cost to the government budget as consumer prices remained unchanged. Thus, it is obvious that any costs of consumer subsidy must be shared by the farmers and the state. Therefore, once the total cost of consumer subsidy is determined, indirect taxation on agriculture (and the cost to the budget) will be determined by the level of the 'scissors difference'. Consequently, the government is able to determine the magnitude of resource transfer from agriculture to the rest of the economy by manipulating the terms of trade against agriculture. Since any increase in the procurement prices of agricultural products means increasing the budgetary cost given a certain amount of consumer subsidy, the government always faces difficulties in making price adjustments.

The recent reforms have greatly increased farmers' incomes and opened many more non-farm employment opportunities. To a large extent, this new economic environment has also pushed up the cost of agricultural production. It is estimated that unit labour costs of grain production increased at an annual rate of 9.4–12 per cent between 1978 and 1985. The cost of current inputs increased much faster at 28 per cent per year over the same period (Grain Research Group of the Economic Policy Research Centre for MOA, 1988, p. 17). Although a significant proportion of the increased input costs was absorbed by improved labour productivity, which increased at 16 per cent annually between 1978 to 1985 (due to organizational and managerial reforms of the agricultural sector), the real cost of grain production increased considerably. Reinforced by expanding demand, the market price for grain increased by 7–10 per cent per year between 1978 and 1988 (ibid).

Because the retail price to urban consumers remained almost unchanged, grain consumption subsidies to the urban population had greatly increased. Wu Shuo estimated that for each kilogramme of grain consumed by the urban consumers the subsidy increased from 0.2 yuan in 1980 to 0.6 yuan in 1988. About two-thirds of this subsidy was paid by the state while the rest was paid by the farmers. Wu also estimated that the state grain subsidy in 1984 alone lost about 23 billion yuan while farmers paid an amount of 10 billion yuan (Wu Shuo, 1988, pp. 15).

Thus the total benefit to the urban consumers in 1984 through grain consumption was more than 30 billion yuan, or 150 yuan per capita, which was more than one-third of the average per capita income of the rural population in that year.

Considering the fact that urban residents in China are much better off than their rural counterparts (see Table 2.2), one may ask why the government should not increase the consumer price in line with the real market price rather than shoulder the increasing budgetary cost and squeeze the farmers. It is difficult to answer this question on the basis of any efficiency and equity arguments, but many scholars may attribute this dilemma to the urban or industrial bias in macro-economic policy pursued by the government.

Poor management of the state-owned industrial sector has also been blamed for failing to increase productivity and efficiency so that it can increase the wages paid to their employees without increasing unit labour costs to absorb the increasing prices of foods and materials.

Economic reforms have brought about rapid agricultural and rural development (township enterprises), but reforms in the state industrial sector have been unsatisfactory.[5] In 1991, 40 per cent of state enterprises were making losses. Although total production increased by more than 10 per cent, total profit and taxes declined by 18 per cent. This has been the general trend over the 1980s. A bigger state budget has increasingly relied on contributions from the collective and private sectors (both urban and rural). The share of state industrial production in total industrial production declined from more than 90 per cent in 1978 to less than 50 by 1991. By contrast, the share of the rural industrial output in total national industrial output increased from 9 per cent in 1978 to 33 by 1991 (personal estimates from *ZGNCJJTJDQ*, 1949–86, and the most recent SSB data).

State subsidies to the state-owned sector were 56 billion yuan in 1991, almost 20 per cent of total state expenditures. Including food subsidies to the urban population, total subsidies to the state sector are equivalent to more than 40 per cent of the total state expenditures.

With an inefficient state industrial sector and an urban biased policy, agriculture can hardly escape exploitation by the state if the government cannot afford and/or does not commit itself to paying increasing costs.

Resource transfers from agriculture to other sectors of the economy may be important for economic development. However, the speed and magnitude of transfers must depend on the ability of agriculture. Unstable agricultural and economic growth in China in the past suggests that an accelerated and excessive resource transfer from agriculture regardless of its ability may incur dangers and disadvantages. These include the following.

(1)   The creation or, more likely, aggravation of rural–urban income inequality conflicting with broader objectives of egalitarian income distribution.
(2)   The creation of 'infant' industries which impose a burden on the economy by continuing to require protection even in the long term.
(3)   Price control to extract resources from the agricultural sector reduces the profitability of some major products such as grain. Reduced profitability not only constrains the producers' incentives, but it may also prevent farmers from exploiting new production opportunities created by advances in science and technology. As a result, breakthroughs in the improvement of agricultural productivity may become much more difficult.
(4)   Stagnation, or even a decline in the amount of agricultural marketed surplus, due to the farmers' reduced purchasing power and the corresponding decline in agriculture's product and market contributions, could eventually lead to industrial decline as well.

5.4.2 Labour contribution

A logical industrialization process should benefit from capital transfers and increased use of cheap agricultural labour. In China, however, industrial development in the pre-reform period did not use much cheap labour from agriculture. Farmers were largely excluded from the industrialization process although they had contributed almost all the accumulated capital assets. This was an inevitable outcome of the state industrial policy characterized by high capital intensity and urban bias.

There were two undesirable consequences of this development strategy. Firstly, labour could not be quickly transferred from agriculture to industries, leading to huge underutilization of human resources and depressed labour productivity of the rural economy. Secondly, industrial labour was highly expensive as enterprises had to employ only the urban workers who were entitled to various state subsidies. However, high labour costs had to be entirely borne by the enterprises. As a result, domestic industrial goods were not competitive and so required heavy protection by the state.

Significant labour transfers from agriculture to industries and other non-farm enterprises were possible during the economic reforms. However, such transfers were confined within the rural economy. This has been characterized by the dramatic development of the RTVEs. Obviously the development of agriculture and non-agricultural enterprises in the rural areas have been mutually beneficial. One great advantage of the rural non-farm enterprises over the state-owned industrial sector is that the former have substantially benefited from using cheap labour released from agriculture. The state sector has 'selfishly' excluded farmers from participation but it has had to bear the inevitable consequences: low profitability – low growth – and eventually self-destruction as competition from the collective and private sectors, particularly from the RTVEs, increases over time.

5.5   FOREIGN EXCHANGE CONTRIBUTION

Agricultural exports have been a predominant source of foreign exchange earnings in China. In the 1950s, they accounted for more than 80 per cent of total foreign exchange earnings. Although by the 1980s their dominant position was gradually supplemented by manufacturing goods, they still contributed about 50 per cent of the total national exports (Table 5.9).

Agriculture's foreign exchange contribution was achieved not only through direct agricultural exports, but also through reducing agricultural imports via increasing domestic agricultural production. Agricultural import substitution may be especially advantageous in certain circumstances. However, there is the question of which kinds of import are technically feasible and economically advantageous to produce at home. But the production of some agricultural products may be expanded at a lower real cost by using reserves of unutilized land and labour. As discussed in Chapter 4, the reason why grain self-sufficiency has been one of the major objectives of agricultural policies in China is closely related to the objective of saving foreign exchange. It is also based on the fact that China still has a comparative advantage in cereal (especially rice) production (see Tables 4.8 and 4.9 in Chapter 4).

The role of agricultural exports is further reflected in their net contribution of foreign exchange earnings. Although a consistent long-term time-series data is not available, the data for the 1982–85 period is sufficient to prove that agriculture has been the most important sector of the economy in terms of net foreign exchange earnings. It also indicates how import substitution can help save foreign exchange expenditures. In 1982–83, when China had to import large quantities of agricultural goods, agricultural imports accounted for almost 40 per cent of the total national imports in 1982 and 27 per cent in 1983. As a result, the net foreign exchange earnings by agriculture were only 12.6 per cent in 1982 and 17.9 per cent in 1983 of the total value of national imports. Good harvests in the following two years helped reduce agricultural imports substantially, to only 19 per cent of the total national imports in 1984 and 12.5 in 1985. Consequently, the net contribution of agricultural exports increased to 24.6 per cent in 1984 and 20.3 per cent in 1985 (Table 5.9).

Table 5.9   Agricultural exports (billion US$ and %)

	Exports		Imports		Export − imports: (1)−(2)	
Year	Value (1)	As % of national total	Value (2)	As % of national total	Value (3)	As % of total national imports
1953	0.83	81.6	—	—	—	—
1960	1.36	73.3	—	—	—	—
1965	1.54	69.1	—	—	—	—
1970	1.68	74.4	—	—	—	—
1975	4.41	60.7	—	—	—	—
1980	8.82	48.3	—	—	—	—
1982	10.06	45.0	7.63	39.6	2.43	12.6
1983	9.63	43.3	5.81	27.2	3.82	17.9
1984	11.94	45.7	5.21	19.0	6.73	24.6
1985	13.88	50.7	5.29	12.5	8.59	20.3

Note: Data for 1953–80 are provided by the import-export departments. Data for 1982–85 are provided by the customs. These two sources of data may not be consistent.
Sources: *ZGNCJJTJDQ*, 1949–86, pp. 517–19 for 1953–80; and pp. 469–76 for 1982–85.

5.6 INCONSISTENCIES IN AGRICULTURE'S ROLE

Although the four contributions of agriculture to economic development are conceptually distinct, they are also interdependent. Thus, these four contributions should be consistent for agriculture to play a real and positive role in economic development. As discussed above, the foreign exchange contribution need not be inconsistent with market and factor contributions. But the capital contribution may not be consistent with other contributions. Colman and Nixson (1986) have pointed out that the 'contributions' of agriculture may be reinterpreted as policy objectives and they list three possible inconsistencies or contradictions which may exist among these four agricultural contributions: (1) the inconsistency between increasing the marketed food surplus and keeping food prices low; (2) the inconsistency between increasing foreign exchange earnings through export crop production and expanding food production; and (3) the inconsistency between taxing agriculture heavily (either through direct taxation or indirect methods) and increasing production of either food or export/industrial crops (Colman and Nixson, 1986, p. 209).

The Chinese government has extracted capital from agriculture by paying low prices for agricultural products and charging high prices for industrial goods sold to the farmers. Low food prices are desirable from the standpoint of keeping down wages in the non-agricultural sector, but they reduce farmers' welfare, marketed surplus and capacity to invest. The greater the priority given to accelerating industrialization, or to public sector growth, the greater the pressure to squeeze agriculture hard, to extract the maximum surplus, and to sacrifice the welfare of farmers.

The same inconsistency exists between food and export crop production. To meet industry's import requirements for capital and intermediate goods, priority has to be assigned to increasing exports and foreign exchange earnings. Increasing agricultural exports entails drawing resources away from food production and raises the problem of extracting an adequate marketed food surplus at reasonable prices. Moreover, increasing agricultural exports may also reduce the availability of domestic supplies. This will heighten the pressure on inflation. The high price inflation in China from 1985 to 1988 may be partly explained by the rapid expansion of agricultural exports in those years. At an important meeting of the Central Party Committee, the then party general secretary Zhao Zhi-yang called for a slow-down in agricultural exports as one of the several methods to combat high inflation (Zhao Zhi-yang, 1988).

Inconsistencies between the agricultural contributions make the formulation and implementation of a policy for agricultural development very difficult. Therefore, before choosing the methods of extracting resources from agriculture and deciding the best rate of transfer, decision-makers need to formulate clear policy objectives. Thus, in order to examine and criticize the Chinese agricultural policies, it is necessary to understand and analyse the objectives of government policy in detail. This is the major theme of the next chapter.

5.7 SUMMARY AND CONCLUSION

This chapter has discussed the four contributions of agriculture to the national economy in China: product contribution; market contribution; factor contribution; and export contribution.

Product contribution of agriculture to the national economy has become less important because the share of agricultural output in total national income has been declining over time. A declining trend of agriculture's product contribution has been due to some inevitable laws of economic development, i.e., the demand for foods and other agricultural products is income inelastic (consumers spend proportionately less of incremental expenditures on agricultural goods); and agriculture has increasingly relied on industries for material inputs (an increasing proportion of agricultural output is accounted for by the non-agricultural sectors).

Government policy discriminating against agriculture has been another important factor depressing the rate of agricultural growth.

Despite the inevitable trend of the declining importance of agriculture, development experience in China suggests that sufficient effort must be made to sustain agricultural growth. The importance of agricultural product contribution is explained in two ways. One is that agriculture provides food for the entire population and input materials for the agriculture-based industrial sectors which account for more than one-third of total industrial output and employment. The other is that agriculture is a large market for domestic industries which produce modern agricultural inputs and consumer goods (the market contribution of agriculture).

Factor contribution had been dominated by capital transfer from agriculture to the industrial sectors before economic reforms. Rural labour was largely excluded from the industrialization process, partly due to state control on rural–urban migration, and partly due to the inability of the state industrial sectors to absorb labour released from agriculture. Significant labour transfers from farm to non-farm activities were possible during economic reforms. Such transfers were restricted to the rural areas, owing to rapid expansion of the rural non-farm enterprises.

Agriculture has been the largest exporting sector in China. The contribution of agricultural export is also reflected in its ability to generate huge net foreign exchange earnings.

The existence of a number of conflicts between the four agricultural contributions has many policy implications. The pre-reform policies focused on capital and foreign exchange contributions of agriculture. Severe capital transfers resulted in a very slow rate of growth in agriculture and stagnation in farmers' income.

Economic reforms have reduced the rate of capital transfers from agriculture and have brought about concomitant higher growth in agriculture and industry. Although the policy changes during economic reforms have not been consistent, the shift of emphasis to agricultural growth has been an obvious element.

A comparison of economic performance before and after the economic reforms in China indicates that a higher rate of capital transfers from agriculture does not necessarily result in higher growth in the rest of the economy. Sustainable economic growth has to depend on a balanced development of agriculture and the industrial sectors.

6 Policy Objectives, Instruments and Mechanisms

6.1 INTRODUCTION

The objectives of Chinese agricultural policies are closely related to the strategy of industrialization. We have argued that agriculture cannot be totally sacrificed to support industrial development. In order to make the four contributions discussed in Chapter 5, agriculture itself needs sustainable and significant growth.

The importance of agriculture as a primary engine of growth has been officially stated throughout the post-independence period. Even during the Great Leap Forward, when agriculture was most severely squeezed to support industrial and urban construction projects, official policy still stated that 'Agriculture is the foundation, and industry is the key sector of the national economy.' This official statement at least reveals that policy-makers were aware of the critical role of agriculture in the national economy, although in practice things turned out to be very different.

What was wrong in the past, especially in the pre-reform period, was that policy objectives were ambiguous and the policy instruments used to achieve them were inconsistent, often resulting in the overtaxing of agriculture and inefficient resource allocation within the rural economy. Increased inefficiency has been exacerbated by mandatory or administrative measures which were not consistent with stated objectives.

Although rural economic reforms have significantly increased the producer prices of all agricultural goods, the government is still able to manipulate the sectoral terms of trade either to reduce or to increase the degree of discrimination against agriculture. Public rural investments can also affect the pace of agricultural development and the living standards of the rural population. However, how much and how fast the government can adjust its policy for or against agriculture depends largely on a thorough understanding of the consequences of policy instruments adopted to achieve the declared objectives.

One difficulty in choosing appropriate policy instruments is that there are many inconsistencies embodied in policy objectives. Instruments can be used to achieve some of the goals but, at the same time, they undermine others. Some trade-offs between different objectives were touched upon in the previous chapters: for example, between industrialization and agricultural growth, and between food and cash crop production.

A complete set of objectives of agricultural policy has never been officially published in China. Government policy documents concerning agriculture and rural development have mostly recorded statements on why and how government would try its best to boost agricultural production and improve the rural living standards. Little has been mentioned about how to tax agriculture more effectively to achieve concomitant development of both farm and non-farm sectors of the economy. Measures used to reduce the financial burden on agriculture have often been adopted passively rather than through orderly planning. Even in the reform period, policy changes favouring agriculture have been possible because the resultant successful performance of agriculture has also strengthened all the other sectors of the economy.

The discussion in Chapter 5 points out the need and inevitable trend to maintain unbalanced growth between agriculture and the non-agricultural sectors. It also points out the need to maintain a sustainable and significant agricultural growth as a precondition for successful industrialization. These two requirements need not necessarily be contradictory if a whole set of government objectives can be clearly spelled out. However, if government policy objectives are ambiguous, future agricultural development will continue to suffer periodic setbacks as experienced in the past.

China shares many similarities with most other developing countries in that agriculture is still the key sector of the economy, and a large proportion of the population is still living in the rural areas. However, some unique social, political and economic characteristics also make China very different from other developing countries in setting policy objectives and adopting policy measures towards agriculture.

Due to the similarities with and differences from other developing countries, the lessons and experiences of agricultural development and rural economic reforms in China may be useful for those countries which are embarking on agricultural reforms, especially for those which used to rely heavily on central planning and strong state intervention.

Section 6.2 is devoted to an understanding of the government objectives and an analysis of inconsistencies between them. Section 6.3 classifies the instruments employed to achieve policy objectives in different groups. The last section outlines the control mechanisms introduced by the state (e.g., setting prices and regulating trade in agricultural commodities).

6.2 POLICY OBJECTIVES

In general, agricultural policies are associated with the objectives of increasing all the four contributions of agriculture. Due to the inconsistencies existing among these four contributions, policy objectives cannot usually be satisfied simultaneously. However, different decision-makers may have different priorities among numerous objectives. Thus, it is important that we understand the policy objectives before any evaluation can be undertaken.

Colman and Young (1990, Chapter 8) point out that any country's policy towards the agricultural sector as a whole or towards one particular interest group such as food consumers, grain producers or fertilizer manufacturers can be characterized as consisting of three sets of elements; (1) objectives; (2) instruments of policy and (3) rules for operating instruments of policy.

A *policy* is usually framed in terms of several simultaneous objectives, and involves several instruments which are applied according to specific rules devised in order to achieve the objectives. It is the way in which the rules are set for the operation of the instruments which determines the outcome of policy, and which thereby controls the extent to which the different objectives are individually achieved. However, what is actually achieved is often substantially different from what is expected by the official policy statements.

Due to the relative importance of domestic agriculture in the national economy, the objectives of agricultural policies in a developing country are quite different from those in a developed country. In a developed country government is more likely to protect the agricultural sector by subsidizing the farmers because the farm sector is relatively small and farmers' living standards usually lag behind those of people engaged in other occupations.

The objectives of agricultural policies in the developed world can be illustrated by those of the European Union. The Union's Common Agricultural Policy (CAP) establishes several objectives for the policy which may be summarized as follows: (a) support to farmers' and farm workers' incomes, (b) increase in agricultural productivity, (c) stabilization of markets, (d) guaranteeing of regular food supplies (which may be interpreted as achieving an unspecified degree of self-sufficiency), and (e) securing reasonable prices for consumers. There are also some other objectives, such as assisting farming and rural communities in more remote and otherwise disadvantaged regions and protecting specific habitats and landscapes. To realize all these objectives, a set of policy instruments must be employed. These include import tariffs and export subsidies, intervention buying to increase domestic prices, production quotas to restrict outputs, deficiency payments, production subsidies and investment grants, plus a range of measures to help dispose of surplus production.

In a developing country agriculture is usually considered a major source of capital. It is generally believed that agriculture is squeezed to support the diversification of the economy. In most developing countries much more emphasis has been placed on keeping food prices down by subsidizing consumers rather than producers, on stimulating agricultural exports to contribute to the balance of payments, and on securing industrial crops such as cotton for local agricultural processing industries.

The objectives of Chinese agricultural policies are very similar to those of many other developing countries in that agriculture and food production have a dominant role in the national economy. But due to some differences at the broader level of the national economy, the objectives of Chinese agricultural policies may also have their own characteristics.

Before discussing the objectives of Chinese agricultural policies, it is therefore necessary to distinguish some fundamental differences between the national policy of China and those of other developing countries.

(1)   Industrialization through rapid development of the heavy-industrial sector has been one of the most important strategies of the Chinese government. Therefore, the objectives of agricultural policies have been directed towards achieving high rates of capital accumulation.

(2)   Like many other developing countries, China has heavily supported a food subsidy programme. However, the real motive for food subsidies in China is not based around equity arguments. Food subsidies are only available to the urban people who are

much better off than their rural counterparts. Due to the long-term urban–rural divide in China, it is not difficult to suggest that the food subsidy programme has the explicit objective of maintaining a certain living standard and ensuring food security to the urban people who are more politically powerful.

Thus, apart from the economic reason that food subsidy is to facilitate the so-called 'low urban wages and high accumulation' strategy (Chapter 1), political and ideological considerations are another reason for maintaining food subsidies in China.[1]

(3)   Unlike other developing countries, the process of industrialization and urbanization in China is unique in that labour migration from the rural areas to the urban areas has been tightly controlled by the government. Rural people are unlikely to find formal jobs in the industrial sector so that they must seek employment opportunities in the local areas.

Although this labour policy has helped China to avoid the problem of serious urban unemployment experienced in many developing countries, it has had a number of negative effects on the development of the national economy. Firstly, it enhances the position of state employees (urban residents) in bargaining for higher wages and state subsidies, leading to excessive labour costs and inertia. Secondly, it encourages the development of capital-intensive industries. Thirdly, it results in enormous underutilization of human resources in the rural areas and stagnation of rural productivity.

As the urban industrial sector is inefficient, the state has not been able significantly to improve the living standards of its employees. To meet the demand for higher wages, food and other subsidies were used as supplements for cash incomes. This may partly explain the dilemma presented in Chapter 4 concerning the government's reluctance to reduce consumer food subsidies in order to reduce budgetary costs and the burden on grain producers.

(4)   With the socialist reform in the 1950s and the collectivization movement in the later decades, major agricultural resources such as land, and production facilities including the irrigation systems have been transferred from the private to the state or the collectives. Such transfers have greatly facilitated the government in squeezing agriculture and controlling the labour movement.

(5)   The huge and expanding population and limited arable land plays a key role in formulating the objectives of Chinese agricultural

policy. Food production and self-sufficiency is particularly important for the Chinese population. Effective utilization of agricultural land is critical in maximising food and industrial crops.

The Chinese government has been widely criticized for not having a clear blueprint for economic development. In the pre-reform period, when grain production was overemphasized, government agricultural policies were broadly defined as 'Grain is a key link, and all the other activities will be developed' (*yi nian wei gan, chuan mei fa zhan*). It meant that grain is the focus of agricultural development although efforts will be made to develop all the other activities.[2]

In the reform period, having realized that it was inappropriate to have overemphasized grain production at the expense of other rural activities, the government modified the new agricultural policy by saying: 'Never slacken grain production, try every possible effort to diversify agriculture' (*jie bu huang song nian se sheng chan, ji ji hua zhan duo zhong jiang yian*). It means that grain production is never neglected while agricultural diversification will be greatly encouraged.

According to these broad statements of policy, it is difficult to find a complete and clear set of objectives, although food self-sufficiency is apparently the top priority. However, based on some scattered materials concerning agricultural development and the discussion in the previous chapters, the objectives of Chinese agricultural policies can still be interpreted as follows.

(1) Agriculture should produce adequate foodgrain to meet the basic needs of farmers and to provide a sufficient amount of surplus for the non-farm population. Food self-sufficiency is a national strategic objective, not just to save scarce foreign exchanges, but also to maintain national food security which is a precondition to ensure social and political stability.

(2) Agriculture should provide adequate raw materials such as cotton, oil-bearing crops and sugar-bearing crops for the rapid development of textiles, food processing, and other agroindustries.

(3) Agriculture should be gradually diversified so that fishery, forestry, animal husbandry and sideline production can be rapidly developed to improve the dietary structure of the population, to stimulate agricultural exports, to increase farmers' incomes, and to widen the scope of rural employment.

(4) Agriculture should maintain stable and low food prices for the urban consumers (involving food subsidies).
(5) Agriculture should contribute as much capital as possible to accelerate industrial development.
(6) Efforts should be made to increase agricultural productivity by using modern technologies combined with traditional ones. It is also important to maintain and gradually improve the ecological conditions of the countryside.
(7) More attention should be paid to the development of the remote and more disadvantaged regions so that inter-regional income disparities can be gradually reduced. The poor regions are more concentrated in the western part of the country where there are more underutilized natural resources, including land. Their main constraints on development are the lack of capital and technology. A more rapid agricultural development of the poorest regions will not only help the local people to improve their living standard; it will also be important for increasing the overall agricultural productivity as there exists a greater potential for agricultural development in these regions due to a more favourable land:man ratio than in the more developed regions.

Of course, there are many conflicts among these objectives. In principle, the above objectives are set to increase the four contributions of agriculture discussed in Chapter 5. The potential conflicts among these objectives are briefly discussed below:

(1) Industrialization and agricultural development

How to balance the development of agriculture and industry has been one of the most difficult problems faced by the policy-makers in China. In the pre-reform period, industrialization was overemphasized and agriculture was largely ignored. As a result, agricultural growth was slow and farmers' incomes stagnated for more than two decades. The policy changes in the reform period to revitalize agriculture have brought about sustainable agricultural growth and improvements in farmers' living standards. Due to strong growth in the rural economy, the state industrial sector and the urban people have also benefited. The state industries and people's incomes in the urban areas have grown faster in the reform period than before. A comparison of economic performance between these two periods clearly indicates that overemphasis on industrialization alone can harm both agriculture and industry since they are interdependent.

(2) Food self-sufficiency and the production of non-food crops

The second conflict exists within agriculture. Diversification of agriculture to improve the production of non-grain crops will inevitably undermine grain production due to limited agricultural resources, especially land. Thus diversification of agriculture is directly contradictory to the objectives of food self-sufficiency and food security.

In China, the pre-reform policies to overemphasize grain production at the expense of non-grain crops did not yield satisfactory growth in all crops because the pre-reform policies undermined the potential benefits of regional specialization in crop production. All regions were forced to produce enough grain for their own consumption regardless of the cost of production. Thus regions with a comparative advantage in producing non-grain crops had to produce grain at high cost while those with comparative advantage in producing grain could not find markets for their products. As a result, the production of non-grain crops was greatly suppressed, reducing the opportunities for farmers to generate cash incomes. Ironically, grain production did not grow quickly either. In contrast, both grain and non-grain crops have been growing faster during economic reforms, although the area sown to grain crops has been declining faster throughout the reform period. This indicates how specialization in production can lead to higher growth under increasing resource constraints.

(3) Urban consumer food subsidies against farmers' incomes and state budgetary costs

Supply of food at low and stable prices is beneficial to the urban consumers, but increased cost of subsidies has to be borne either by the state, or the farmers, or both. Increased state budgetary cost leads to reduced investments in other areas of the economy, while depressed farmers' income leads to lower incentives of production.

China has successfully abolished the subsidies for all the non-grain food items since 1985, but grain subsidies still remain untouched except in some experimental regions where the local governments are trying to reduce or abolish grain subsidies. In most regions, the conflict between the need to stimulate grain production and the need to maintain grain subsidies has been and will continue to be the most difficult problem faced by the government, although many studies have indicated that urban grain subsidies can be abolished without much difficulty (see Chapter 4).

The abolition of grain subsidies may have to be compensated for by higher wages in the urban sector. This means that the state may have to incur more budgetary costs. However, it is likely that farmers will respond quickly to higher market prices by producing more. Increased production will, in turn, suppress market prices. As a result, urban consumers may not suffer as much as it may appear if grain subsidies are abolished.

(4) Agricultural development and environment

Another conflict is found to exist between agricultural development and improvement in the ecological system. Maintaining or improving the present ecological system needs more resources to maintain or increase the coverage of forests throughout the country. This may reduce the resource availability for the crop-farming sector. In addition, improvement in land productivity, which is the major means of increasing agricultural output in China due to land availability, has greatly depended on the heavier use of chemicals. Thus rapid agricultural growth without paying much attention to the environment will inevitably damage the ecological system. This, in turn, will accelerate land erosion and would have a detrimental effect on agriculture in the long term.

(5) Regional development strategies

Although the development of the poor areas is beneficial to the whole country in the long term, significant growth in these areas may require substantial investment to improve the infrastructure, especially the irrigation and transportation systems. Huge investment in these areas will slow down the development of the rich regions. To maintain an adequate agricultural growth rate in the short run, however, investment in the rich regions cannot be neglected because they have a much bigger share in agricultural growth than the poor and remote areas.

6.3 POLICY INSTRUMENTS

The objectives discussed in the last section can be realized through policy instruments. In general, there must be at least as many instruments as there are objectives, but there may be several alternative policy instruments to achieve the same objective. For instance, if the objective is to increase production, the instrument chosen might be a production subsidy, an import tax, an input subsidy or a capital grant. If the objective is to subsidize the consumers, the instrument chosen might be a consumer subsidy in terms of lower prices or a lump-sum payment. Because there are so many policy instruments, some authors have tried to classify them into distinct categories. For example, McCalla and Josling adopt a classification based on five levels: (1) frontier; (2) consumption or retail level; (3) product market; (4) input markets and (5) fixed factors (McCalla and Josling, 1985, pp. 108–9). Colman and Young use a different approach by considering the level in the production and the distribution system at which intervention is applied. Using their approach, instruments are listed according to whether they are imposed (1) directly at the farm level, or (2) at the national frontier, or (3) at some other point in the domestic market (Colman and Young, 1990, Chapter 8).

The policy instruments which have been used by the Chinese government to achieve policy objectives are presented in Table 6.1a and simple explanations of these instruments in Table 6.1b. It is worth pointing out that these instruments are all relevant to the developing countries. Some instruments which are exclusively used by the industrialized countries (typically, the EU, USA and Japan), such as deficiency payments, land set-aside schemes and subsidized exports are not included in the table.

Table 6.1a   Classification of selected policy instruments

Farm-gate:

(1)   Production subsidies/taxation
(2)   Input/subsidies/credit
(3)   Production or acreage quotas
(4)   Compulsory food procurement
(5)   Land reform measures

Market:

(6)   Parastatal trading and marketing boards; price discriminations
(7)   Intervention buying and public stock management
(8)   Food subsidies to consumers
(9)   Public investment in infrastructure, education and research

Frontier:

(10)   Import tariffs, levies or duties
(11)   Export subsidies or taxes
(12)   Non-tariff barriers

Table 6.1b   Definitions of selected policy instruments

Farm-gate level:

1.   Production subsidy/taxation: A fixed or proportionate subsidy (product tax) paid (charged) per unit of output.

2.   Input subsidies/credit: Subsidies per unit of input. Cheap credit offered for the purchase of inputs will have the same effect.

3.   Production or acreage quotas: Where total production or acreage of a crop is imposed, individual farms have to commit a minimum amount of land for production of the specified crop, be it a cash or food crop.

4.   Compulsory food procurement: Producers may be required to sell minimum quantities of grain to state trading organizations at below market prices.

5.   Land reform measures: Legislative measures may be enacted to control landlords and tenants rights, or to reallocate land rights. Payments may be offered to promote land amalgamation or to encourage older farmers to retire.

Market levels:

6.   Parastatal trading or marketing boards: The state may authorize the creation of commodity trading bodies with a variety of powers. They may be constituted as monopolies or monopsonies to exercise market power in a variety of ways, e.g., increase producer prices by discriminating monopoly prices, tax producers by applying monopsony powers.

7.   Intervention buying and public stock management: A parastatal organization may be empowered to help place a floor price in the wholesale market by purchasing a commodity at a pre-announced 'intervention price'. The marketing board may also be asked to hold the *strategy grain reserve*, and even a *buffer stock* to stabilize market prices.

8.   Food subsidies to consumers: Parastatal organizations may be used to manage the distribution of low price basic food supplies to consumers. Subsidies are required to finance the gap between the prices at which these organizations secure supplies and the lower prices charged to consumers.

9.   Public investment in infrastructure, education and research: Public sector investment in physical and human capital stimulates economic activity at all stages of the distribution chain, by making available the services or products of capital (roads, research finding, trained manpower) at no direct cost to the farmers.

Frontier levels:

10.   Import tariffs, levies or duties: Taxes on imports may be charged in several ways. They may be as a fixed sum per unit, as a fixed proportion of the value, or as a varying sum equal, say, to the difference between a fixed minimum import and a variable international price.

11.   Export subsidies or taxes: As the counterpart to import taxes, exports may be promoted by fixed, proportional or variable subsidies. In some cases, however, exports have been taxed to discourage them.

12.   Non-tariff barriers: A large number of legislated instruments may be used to impede imports. Heath regulations, labelling requirements, and special technical requirements, which may be continuously changed, are all used to restrict imports.

It is worth noting that different policy instruments might share the same objective, but the effects or the consequences of different instruments may not be the same. Due to these different effects, the choice of instruments to achieve an objective may change under different social and economic conditions. It may become more difficult to choose an 'appropriate' instrument when the losses of social welfare of different instruments are not equally shared by different regions or groups of people.

To demonstrate the complexities of policy instruments and their effects on various aspects of the economy and the interests of different groups of people, Table 6.2 provides in summary internal effects of five policy instruments which are most commonly used in the developing countries: (1) input subsidies for the producers; (2) consumer subsidies; (3) production taxation; (4) export taxation and (5) import taxation.

The main costs and benefits of these policies can be summarized as follows.

(1) Input subsidies

In pre-reform period, input subsidy was used as partial compensation to output taxation. For instance, farmers usually received an assured amount of chemical fertilizers or insecticides at fixed and low prices from the rural supply and marketing co-operatives when they delivered grains and/or industrial crops to the state procurement agencies. Farmers also received subsidies for the use of other farm inputs, such as electricity and machinery. Agricultural input subsidies have been abolished since 1985 in an effort to reform the rural price system as many non-cereal agricultural products were allowed to be traded in the free markets.

The benefits of input subsidies include increased domestic production, improved balance of payments due to increased self-sufficiency or more surplus for export, and increased income to producers. The disadvantage of input subsidies is increased budgetary cost to the government.

(2) Consumer subsidy

Consumer food subsidy is probably the most important policy instrument relating to Chinese agriculture.

The benefits of consumer subsidies include price stability and increased income to consumers. The negative effects include higher

Theoretical Analyses of Agricultural Policies

Table 6.2   Effects of selected policy instruments[a]

	Input Subsidy	Food Subsidy	Prod. Tax	Export tax	Import tax
Domestic price effects					
Producer price	[d]	N.A.	—	—	+
Wholesale price	[d]	—	N.A.	—	+
Retail price	[d]	—	N.A.	—	+
Production effects					
Output	+	N.A.	—	—	+
Quantity of input	+	N.A.	—	—	+
Price of inputs	[e]	N.A.	N.A.	N.A.	N.A.
Consumption effects					
Consumption	[d]	+	N.A.	+	—
Trade effects					
Net imports[b]	—	+	+	+	—
Balance of payments[c]	+	—	—	—	+
Public expenditure effect					
Budgetary cost	+	+	—	—	—
Tax revenue	N.A.	N.A.	+	+	+
Redistribution effect					
Producer surplus	+	N.A.	—	—	+
Consumer surplus	N.A.	+	N.A.	+	—
Taxpayer cost	+	+	—	—	—

Notes:
[a] A positive sign denotes an increase and a negative sign denotes a decrease.
[b] A negative (positive) sign signifies a decrease (increase) in imports (exports).
[c] A positive (negative) sign denotes an improvement (worsening) of the balance of payments.
[d] If domestic price is in alignment with the parity prices, input subsidy would not change the domestic commodity prices. However, if the country is a close economy, i.e., the domestic price is determined by the domestic supply and demand, then an input subsidy will lower the domestic producer and wholesale as well as retail prices, thus domestic consumption will increase as a result.
[e] The input price to farmers will fall, but input suppliers may charge higher prices than before the subsidy because of increased input demand. Both are simultaneously possible because of the subsidy – in other words, input suppliers may capture part of the subsidy.

budgetary costs and the worsening of the balance of payments due to increased imports.

(3) Production taxation

In China production taxation on agriculture centres around the indirect resource transfers discussed in Chapter 5. It implies that farmers receive low prices for their products delivered to the state.

The only benefit of production taxation is a budgetary gain. The negative effects include depressed domestic production due to low prices and incentives for producers; a worsening of the balance of payments due to low domestic production; and the deterioration of producer welfare.

(4) Export taxation

Export taxation on agricultural products has been realized through currency overvaluation and direct tariffs. As agriculture is a large net exporter in China, currency overvaluation has a significant impact on agricultural exports and farmers' incomes. Before economic reforms, the Chinese yuan was pegged to the US dollar at a fixed exchange rate which implied significant overvaluation. The yuan has been gradually devalued since 1983. By 1990, the official exchange rate was very close to the free market rate, implying an insignificant degree of over-valuation. Thus the extent of export taxation on agricultural exports through currency overvaluation must have declined considerably during economic reforms.

Export taxation depresses domestic prices, benefits domestic con-sumers and increases the budgetary revenue. However, it discourages production for export, worsens the balance of payments and depresses producers' welfare.

(5) Import taxation

Import taxation in China has been mainly imposed on industrial goods (including agricultural inputs and consumer goods). It is generally intended to protect domestic industries. Import taxation raises domestic prices, encourages domestic production, improves produ-cers' welfare, increases government revenue and improves the balance of payments. However, the benefits of import taxation mainly accrue to the industrial sector and government. Agriculture and rural consumers are net losers as import taxation inflates domestic prices of agricultural inputs and consumer goods.

6.4   POLICY CONTROL MECHANISMS IN CHINA

In relation to how policy instruments are implemented, it is useful to make a distinction between price and non-price instruments. Price instruments, such as consumer subsidies, indirect production taxation and foreign exchange control, are usually manipulated to alter the terms of trade for or against agriculture. In China, the whole price system comprising many price instruments has been designed to tax agriculture, directly or indirectly. Thus, price instruments can also be regarded as the 'primary' tools to achieve policy objectives. Non-price (or quantitative) instruments are always imposed as supplementary tools for price control. For instance, consumption rationing is a method supplementary to consumer subsidies. Production quota or compulsory procurement is imposed together with indirect taxation of production.

The objective of price control is relatively easy to understand, but that of quantitative control is not straightforward due to its secondary importance in achieving policy objectives. However, the critical role of quantitative methods cannot be underestimated for two reasons. Firstly, quantitative control can enhance the effectiveness of price control. For instance, if government procurement prices for agricultural products are too low, farmers lose incentives to sell their products to the state. As a result, indirect taxation through low procurement prices will become ineffective. Secondly, quantitative methods help reduce the social cost of price policy. In the case of consumer subsidies, if rationing is not imposed, consumers tend to consume more food at subsidized prices than at free market prices. With rationing, total demand for food can be restricted. Rationing not only helps to reduce the budgetary cost, but also the so-called 'dead-weight' or efficiency loss resulting from excessive consumption induced by subsidized prices. The same argument is valid for compulsory production quota when it is imposed as a supplementary tool for indirect taxation of production. Indirect agricultural taxation through low procurement prices tends to discourage production, resulting in lower domestic production and a dead-weight efficiency loss. With a compulsory production quota, the losses in domestic output and efficiency can be mitigated (for a mathematical solution of this, see Part III).

Government agricultural policies not only involve the choice of instruments, they also need to be implemented through a range of institutions. In most developing countries, including China, government policies are usually implemented by a number of state parastatals who are empowered with monopolistic authority in marketing (Yao and Hay, 1991). These state marketing agencies are able to control prices, quantities, imports, exports, transportation, storage, foreign exchange, and agricultural investments.[3]

This section discusses in detail how the Chinese government intervenes in grain procurement and distribution to achieve its policy objectives.

The first step of government intervention in grain marketing is through setting up an institutional framework in which prices and quantities are determined and/or regulated. In China, this institutional framework has been designed in such a way that price and trade regulations are separately conducted by different government departments and marketing authorities.

6.4.1 Price control mechanism

Two sub-ministerial bureaux, the State General Price Bureau (SGPB) and the State Administration of Industry and Commerce (SAIC), directly under the State Council (SC), are responsible for setting and regulating prices of agricultural products.

The price-setting and monitoring system is illustrated in Figure 6.1. In consultation with a number of ministries, including the ministries of agriculture (MOA), finance (MOF), commerce (MOC) and the State Planning Commission (SPC), the SGPB is responsible for setting and adjusting the planned (contract) prices and negotiated prices of agricultural products. These contract and negotiated prices are implemented at the provincial and local levels by the price bureaux at the corresponding levels. As the SGPB enjoys a sub-ministerial status, all the local branches have the same power at the respective levels of local government.

Employing its comprehensive system throughout the country, the price bureaux monitor official procurement and sales at all market

Figure 6.1   The price-setting mechanism of agricultural marketing in China

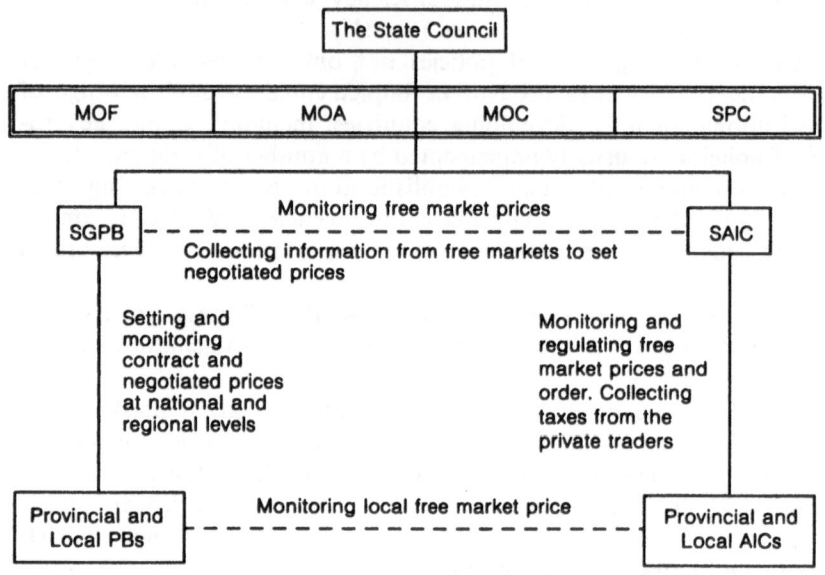

Notes:
MOA: Ministry of Agriculture; MOF: Ministry of Finance; MOC: Ministry of Commence; SPC: State Planning Commission; SGPB: State General Price Bureau; SAIC: State Administration of Industry and Commerce. Solid lines indicate direction of control. Dotted lines indicate exchanges of information.

levels to ensure compliance of national price policy. The price bureaux also monitor the movements of free market prices for two reasons: one is to ensure spatial and inter-seasonal price stability in the free markets; the other is to collect information for future price adjustments.

The SAIC enjoys the same status of power as the SGPB and has a similar network throughout the country. The SAIC does not have the power to set prices but it is authorized to monitor and, if necessary, regulate local free market prices. The local AICs are responsible for issuing licences, collecting taxes, maintaining order, setting regulations, collecting and disseminating market information including prices. The AICs own and regulate most of China's urban and rural free markets, the number of which increased from 40,000 in 1980 to more than 70,000 in 1988. They collect a 1 per cent management fee and a 2 per cent turn-over tax on free market sales. These revenues are used to offset the network's local operating costs and, in combination with local government and private funds, as investment funds for the expansion of free markets and associated infrastructure.

This two-tier system of price and market control was initially created to ensure the effectiveness of state price policy which discriminated highly against agriculture. The SGPB was empowered to set and regulate agricultural prices at the expense of farmers to benefit urban consumers. The SAIC was in effect a 'watch dog' in preventing free marketing. The local AICs set regulations to ban inter-regional trades and local transactions in large quantities. Thus, farmers would have to deliver most of their products to the state procurement agencies, i.e., the Grain Bureaux of the Ministry of Commerce.

As large-scale free marketing activities (in terms of quantities and number of products allowed to be traded in the free markets) have been encouraged since economic reforms, the roles of the SGPB and the SAIC are greatly modified to cater for the needs of the new market environment. As a matter of fact, the SAIC nowadays has a strong incentive to encourage free marketing instead of being a 'watch dog' as it was in the past because its revenue is greatly determined by the volume of transaction in the free markets. The SGPB has also played an increasingly important role in discovering and formulating free market prices through communication with the SAIC and frequent adjustments of negotiated prices.

6.4.2 Trade control mechanism

Domestic trade

In the pre-reform period, the state marketing system procured and distributed as many as 230 agricultural products and controlled 85 per cent of China's total marketed agricultural production. Since the introduction of economic reforms, the state marketing system has sharply reduced its share of trade in most commodities except grain, cotton and a limited number of other industrial crops.

The MOC has been responsible for procurement, storage, processing, transport and distribution of all these controlled commodities.[4] Despite its commercial nature, like all the other governmental departments, the MOC has regional bureaux throughout the country down to the county level. The MOC's commercial system is also associated with the provincial and local level Supply and Marketing Cooperatives (SMCs). The SMCs supply to the farmers most of their agricultural inputs and some consumer goods. The SMCs were initially created as rural trading co-operatives owned by farmers' collectives. During the Great Leap Forward, these co-operatives were brought under direct state control. Before the economic reforms the ownership and control switched many times between the state and collectives. These SMCs are now considered 'non-state owned', but they are still subject to considerable control by the MOC (the state trading arm).

Domestic trade of grain can be illustrated in Figure 6.2. At the farm-gate level, farm inputs are controlled by the MOC and SMC. In the pre-reform period, agricultural inputs were sold to the farmers at subsidized prices as partial compensation to output taxation. Since 1985, input subsidies have been totally abolished.

Farmers have three major marketing outlets for sales: contract sales at contract prices (contract is mandatory); above-contract sales at negotiated prices; and sales to the free market at free market prices.

Total marketed surplus including all mandatory (contract), nego-tiated and free market sales accounted for 20–26 per cent of grain production during 1952–1982; it increased to 30–35 per cent between 1983 and 1988. State procurement, both mandatory and negotiated, increased from 80 per cent of the total marketed surplus in 1952 to 100 per cent during the Cultural Revolution, and has declined to about 85 per cent at present.

During the economic reforms, state procurement of grain at negotiated prices increased rapidly from 15 per cent of total state procurement in 1980 to 41 per cent in 1987 (Table 6.3). The share of total marketed surplus transacted through urban and rural markets increased from 7 per cent in 1980 to 15 per cent in 1987.

Grain distribution was entirely controlled by the state marketing system before 1980. By 1987, about 7 per cent of total sales was transacted through the free markets. The MOC's Grain Bureaux sold grain to both urban and rural consumers. In the 1980s, about one-third of the total state grain distribution was sold back to the rural areas at negotiated prices. This indicated an increased degree of commercializa-tion and production specialization in the rural economy. Before 1980, all the state sales to urban consumers were at subsidized prices. Since 1980, part of state sales to urban consumers is at negotiated prices. The proportion of negotiated sales by the state agencies to urban consumers increased from 7.7 per cent in 1980 to 28.2 per cent in 1987. This may

Figure 6.2   The grain marketing structure in China

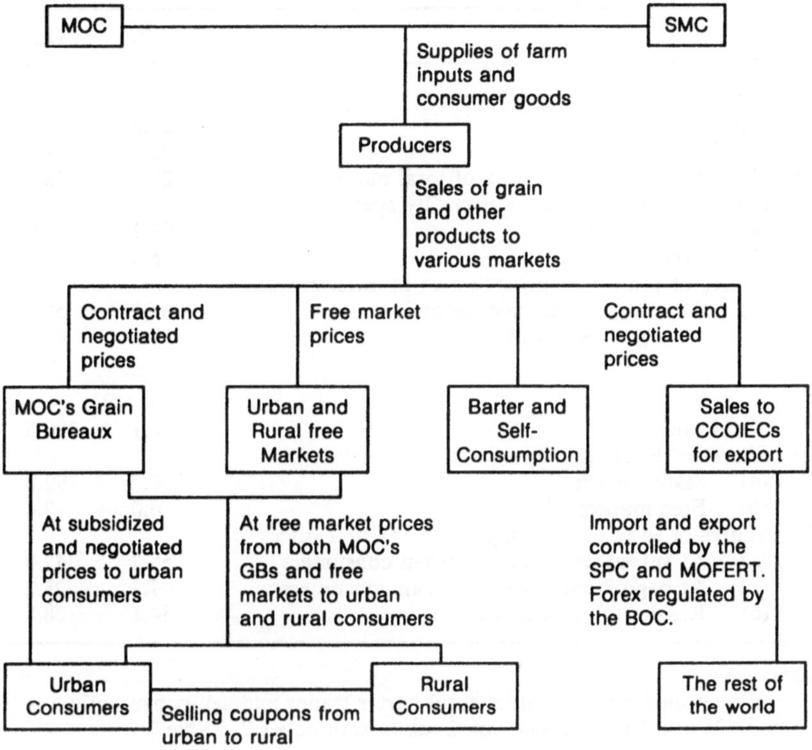

Notes:
SMC = Supply and Marketing Cooperatives. GB = grain bureau. MOFERT =
Ministry of Foreign Economic Relations and Trade. CCOIECs = Cereal, Oil
Import and Export Corporations. BOC = Bank of China.

imply that the government has been trying to reduce consumer food
subsidies in recent years. It may also imply that there is an increasing
proportion of urban people who are new emigrants from the rural
areas and are not entitled to state subsidies. Whichever is the main
reason, the fact that an increasing proportion of the total state sales is
at negotiated prices suggests that it will be much easier for the state to
abolish consumer food subsidies totally in the future.

Table 6.3   Marketing of grain by market (million tons and %)

	1980	1987
Procurement		
(1) Total production	320.6	403.0
(2) Total marketed grain[a]	73.0	141.2
(3) Total marketed grain as % of total output	23.0	35.0
(4) Shares of total marketed grains by agents:		
(a) State agencies	93.1	85.0
(b) Free markets	6.9	15.0
(5) Shares of total state procurement at different prices		
(a) At contract (planned) price	85.1	58.5
(b) At negotiated price	14.9	41.5
Sales		
(1) Total sales	76.1	120.5
(2) Shares of total sales by agents		
(a) State agencies	100.0	92.6
(b) Free markets	0.0	7.4
(3) Shares of sales by state agents at different prices		
(a) At subsidized prices to urban consumers	58.2	43.0
(b) At negotiated prices to urban consumers	7.7	28.2
(c) Resales to rural areas	34.1	28.8

Note:
[a] Total marketed grain excludes rural barter trades and self-consumption.
Source: World Bank (1991), vol. 2, discussion paper 7, Table 1.

International trade

International trade in agricultural commodities is part of China's national economic planning. The SPC sets preliminary and long-term targets for broad categories of imports and exports.

Imports and exports are monopolized by the Ministry of Foreign Economic Relations and Trade (MOFERT), which was established in 1982 with the amalgamation of the Ministry of Foreign Trade, the Ministry of Foreign Economic Relations, the Foreign Investment Commission and the State Import–Export Commission. The MOFERT has its branches in all provincial and municipal governments and 15 product-specific national foreign trade corporations (FTCs). The FTCs exercise monopolistic power in international trade for the respective commodities under their control. They operate on a commercial basis and are responsible for their own profits and losses subject to state foreign exchange regulations.

The China National Cereals, Oils and Foodstuffs Import and Export Corporation (CEROILS) controls most of China's international trade in grain and vegetable oil.

6.5 SUMMARY

As the objectives of the Chinese agricultural policies have been ambiguous and often contradictory, it is important to examine these objectives according to scattered information and the development experience over time. This chapter puts together a whole set of objectives which reflect what the government had intended to achieve but failed so to do in the pre-reform period. The reasons for failure are complicated, but ambiguity of policy objectives and the conflicts between them are among the most important ones. In China, the major conflicts are between industrialization and agricultural growth and between food and non-food production within agriculture. The pre-reform policies appeared to have overemphasized the needs of industrial development and food production but neglected the importance of agricultural production in general and agricultural diversification in particular. A comparison of development experiences before and during the economic reforms provides a good basis for establishing a set of objectives and to understand the potential conflicts.

To achieve the objectives, a set of policy instruments are required. In general, there are as many policy instruments as objectives. However, the fact that each objective can be achieved by a number of alternative instruments suggests that a right choice of instrument at the right moment is particularly important. It is suggested that a combination of price and quantitative instruments can considerably reduce the negative effects induced by price instruments alone.

Like other countries, China has implemented its policies through a complicated administrative system. In particular, price and trade regulations are conducted by different administrative systems. Prices are determined by the central government via the SGPB and the SAIC. Trade regulations are the responsibility of the MOC (domestic) and the MOFERT (foreign). The separation of price and trade regulations may create some difficult problems in food marketing, especially when different authorities have their own interests which are not compatible with each other. However, it is encouraging that there is a greater degree of commercialization and an increasing proportion of the total state sales of food transacted at negotiated prices in recent years. This provides a good opportunity for further market reforms, especially the total abolition of consumer food subsidies and the strengthening of regional production specialization. This will be the main theme of discussion in the third part of the book.

7 Empirical Results and Future Grain Policies

This chapter presents the summary results of the empirical study presented in the Appendix and identifies a few important points for the making of future grain policy in China.

The results of the empirical study demonstrate the effects of six different policy scenarios. They are obtained through applying a quadratic spatial equilibrium model using the cross-sectional data of 1985.

The policy simulation models are constructed based on the theoretical analysis made in Chapters 5 and 6. With the cost–benefit analysis approach, we are able to compare the advantages and disadvantages of different policies as opposed to an entirely free market system. The main policy instruments include production taxation, consumer subsidy and production quota. These three instruments are the most important tools used by the Chinese government throughout the post-independence period. Although market reforms since the late 1970s have greatly mitigated the effects of these policy instruments, they are still very important policy issues whose implementation has significant effects on grain production and farmers' incentives.

One obvious advantage of the quantitative study presented in the Appendix is that it examines the sensitivity of grain production and efficiency with respect to various policy combinations. Due to inevitable technical and theoretical limitations of the models, the figures generated do not infer any serious accuracy. However, they do point out the general direction and degree of distortion caused by alternative methods of government intervention.

The second advantage of quantitative study is that it enables us to examine in detail the possible effects of policy at a regional level. This would not be possible if we just used a geometric analysis like that usually shown in a textbook. Like many other developing countries, China is confronted with the problem of acute unbalanced regional development. This is because the economic boom of the 1980s has been accompanied by an increasing gap between regions. For a country to have a sustainable growth with minimum internal instability, more

171

effort should be made to promote production in the less developed areas. Government support in agriculture and grain production is an effective way to achieve this objective as most poor areas are primarily agrarian. Thus it is important to understand which policy can be used effectively for this purpose.

Another advantage of quantitative analysis is that it enables us to examine the effects of more than one policy tool simultaneously. A quantitative approach is able to calculate the combined effects of a number of policy tools under a certain set of constraints. It can also identify related factors causing specific changes which are not easily recognized without rigorous numerical analysis.

Like many other quantitative studies, the quantitative exercise in the Appendix has its own limitations. To fully appreciate the usefulness of the models, these limitations have to be discussed before the computational results and policy implications are presented. Thus, the structure of this chapter is as follows. The first section (7.1) will discuss the limitations of the quantitative approach in the Appendix (Chapters A2 to A4). Section 7.2 summarizes the computational results estimated from the models. The discussion of computational results then leads to three important comments on grain policy. Section 7.3 explores again the more technical issues of how models can be further developed and utilized for policy analysis and evaluation in the future. The last section identifies three important issues concerning the future direction of grain policy.

7.1 LIMITATIONS OF QUANTITATIVE ANALYSIS

Chapter A2 presents six policy models for the Chinese grain sector. Each model has the same format of quadratic spatial equilibrium with a set of assumptions. Some of the assumptions are common to any other similar studies but others are unique in our case. These unique assumptions are required mainly because available data is inadequate.

Data sources for the study include the *China Statistical Yearbook* and the *China Agricultural Yearbook* published by the SSB. These two series of books provided the major time series data utilized in the models. There are some other data sources. For example, the grain prices and transportation costs were collected directly from the State Price Bureau. The State Bureau of Industrial and Commercial Administration, the MOC and the MOA have much more information and data for the market prices of farm crops. This latter data, however, can hardly be traced in the statistical yearbooks, although the author succeeded in finding some grain market price data for 1987 directly from the State Price Bureau of China.

One problem about data is consistency. As data is collected from different sources of publication, it is very difficult to obtain consistent data series over time as different publishing authorities do not have a uniform way of calculation. The problem is particularly serious with respect to price data. Data on market prices is scattered and does not appear in official statistics. Data has to be collected from various relevant authorities. This then becomes one of the most difficult problems in any empirical study.

The second problem about data is quality. Although the reliability and accuracy of official data (the statistical yearbooks mentioned above) has greatly improved over the last 15 years, there are still defects in the data collection and processing system.

One of these defects is the lack of official publications over time. This makes it impossible to collect consistent time-series data of many important items for meaningful econometric analysis. The SSB only started printing the *Statistical Yearbook* and *Agricultural Yearbook* in 1981. Corresponding data for the earlier period between 1949 and 1980 are either inconsistent or discontinuous. Some earlier statistical yearbooks, such as the volume called 'The Great Ten Years', are highly unreliable. Many important statistical indicators were greatly exaggerated and, therefore, unrealistic. Data published by some United Nations organizations, such as FAO, and those estimated by some western specialists on China may not be consistent with the official data and may have their own measurement errors. Under such circumstances, the choice among different data sources should be made with great caution in order to compile realistic and consistent data series.

Another defect of the system is that the data concerned with prices and incomes is highly inadequate. In this study, one major difficulty was the collection of the price data of different farm crops and per capita disposable incomes at a regional level. Direct data on prices of farm crops are almost non-existent in major statistical yearbooks. This was the main reason why it has not been possible to estimate the 'desired' regional demand and supply functions.

Prices at the national level used in our study to establish the national per capita grain demand function were derived from the price data of grain supplied by the SSB, deflated by the price index for farm products. Grain prices are the weighted average prices of different types of grain, including rice, wheat, corn, soybeans, tubers (which are converted into wheat equivalent at a ratio of 5:1) and other miscellaneous crops. Prices for individual crops, however, are virtually non-existent. This has been the most important obstacle to establishing the demand and supply functions of individual crops. Consequently, it is not possible to obtain any cross price elasticities.

The last data problem is aggregation. Due to the above defects, it is impossible to establish a spatial equilibrium model to estimate the interdependent effects of different farm crops in terms of acreage or prices. To overcome the data problem, only a single-product spatial equilibrium model in which a number of different grains are aggregated into a single product is constructed. Such an aggregation not only makes it impossible to test the substitution effects of different grains for consumption, but also makes it impossible to describe the production patterns of individual crops. Ideally, it is much better to analyse the production patterns by crop because different regions may have different comparative advantages in the production of one, or several particular crops. Even within the same region, the response to a price change of a particular commodity (such as rice) may vary with different grain crops. For example, if all the procurement prices of all types of grain are increased by a certain percentage, the responses of acreage or production of different crops may not be exactly the same. One assumption which could justify the aggregation of different crops is that all these crops are homogeneous in terms of both supply and demand. This explicitly implies that the response of production or acreage and demand for different products must be the same with respect to a price (or income) change. The assumption of homogeneity of different crops leads to an inevitable aggregation problem.

The problem of aggregation is more obvious when the price data are compiled. China has been using the two-tier (or sometimes three-tier) price system for many industrial and agricultural products. When the models are constructed, two different prices need to be considered: the state procurement price and the free market price. Before 1984, the procurement price of grain had two different fixed levels, i.e., the basic procurement price and the premium price. These two prices were replaced by a single contract price in 1985 (see Chapter 4).

On the demand side, consumers are divided into two different groups, i.e., the urban consumers and the rural consumers. These different groups of people not only have different income levels but also face different consumer prices.

In general, consumers in the urban areas have a much higher income than their rural counterparts. The urban consumer price is set by the state much lower than the market price. Consumers in the rural areas buy grains from the free market at the market price.

Grains retained for home consumption by producers do not have a market price. In the model, however, this type of consumption must also be given a value for computational purpose. If the consumption of commercial grains is priced by its market value, how is home consumption priced? Pricing home consumption remains a controversial subject in the literature. Grain producers may put more emphasis on subsistence need and security than on the market values of their products.[1]

The need to ensure subsistence and food security may also vary among different income groups of grain producers. In theory, the high-income groups are more likely to accept risk and pay less attention to subsistence and security than the low-income groups. Thus, different groups of farmers may put different values on home consumption. The difference of risk-averse behaviour among different income groups has important implications in establishing a spatial equilibrium model. In China, farmers' incomes vary substantially across different rural areas (Lardy, 1983; Yao, 1987), and so does the risk-averse behaviour of farmers in different regions. Thus the price placed on home consumption by producers in different regions cannot be the same.

As it is almost impossible to make this kind of distinction, we have to assume that home consumption has the same value as commercial grain and is valued at the market price.

The compilation of data for transportation costs also has some problems. It is not possible to find any detailed inter-regional transportation costs from the statistical yearbooks mentioned above. The Ministry of Railways or the SSB publish a handbook every year in which there are standard freight charges in terms of the cost per ton-mileage according to different types of distance range. The charges listed in the handbook do not cover the costs of storage, waste, loading and unloading. To estimate the real inter-regional transportation costs, we have to rely on some subjective judgement. In addition, given a certain trade volume, the transportation costs between regions may vary significantly if the compositions of total flow carried out by different means of transportation, i.e., railways, waterways and high-ways, are different. Indeed, different regions may have different compositions of these means, but we do not know exactly what the compositions are for individual regions and individual routes between regions according to the data available. Furthermore, one may argue that the standard charges set artificially by the government may not reflect the true market price of transportation because the Ministry of Railways may have monopoly power in setting the charges.

Although efforts have been made to consider this problem, the estimations made with some subjective judgements may well incur errors.

The Chinese transportation system is one of the major bottle-necks in the economy. It is also a major barrier to grain marketing and production. However, in our models, we do not consider this issue due to data shortage. In other words, it is assumed that there is an efficient and adequate transportation network and grain can be freely trans-ported and marketed between regions. Thus, the optimal production patterns derived from our models would look somewhat different from the real situation.

Another artificial assumption about transportation costs is that all the intra-regional transportation costs are ignored. Thus the farm-gate price and the local consumer price is exactly identical within the same region. The choice of supply and demand centres for each individual region is also subjective. This is because the centre of grain supply in a particular region may not be exactly the centre of grain demand. Since there is no corresponding data available at a sub-regional level, it is impossible to estimate the appropriate demand and supply centres separately so that we can consider the intra-regional transportation costs. In reality, intra-regional transportation costs are important because each region in the model covers a huge area.

Another obvious limitation of the models is their static nature. Policy models in this study are based on 1985 data. Computational results are strictly appropriate only for that particular year. If we want to test the effects of policy changes on the production patterns for some other years, the data base has to be changed. Therefore, the equilibrium model can only be used as a static policy simulation model, not as a forecasting tool.

Of course, regional demand and supply functions of grain must be changed as the national economy develops. In the last fifteen years, rapid economic development in China has been accompanied by increasing unbalanced development among regions. The eastern and coastal areas have been growing much faster than the rest of the country. Per capita disposable income has also been widening between the coastal and inland regions. This increasing income gulf will have different effects on grain demand in different parts of the country.

Thus, regional demand patterns and income elasticities of demand may have changed over time.

Unbalanced regional development may also shift the regional comparative advantage pattern for grain production. Regions with higher economic growth rates may lose their comparative advantage in agriculture and grain production. Farmers in the more developed areas have more opportunities to earn income outside agriculture than their counterparts in the relatively poor and more agrarian regions. As more farmers move out of agricultural activities, the wage rates in the agricultural sector will go up, leading to inflated production costs. Thus regional comparative advantage in agriculture fades over time in the more developed areas. Due to its 'static' nature, unfortunately, such income growth effects are difficult to incorporate into the models.

Another shortcoming of the model specification is that it is assumed that the country is a closed economy and entirely self-sufficient in grain. Although grain import and export consist of a tiny proportion only of the country's total output, the importance of trading grain with the rest of the world cannot be entirely ignored. This issue will become increasingly important should China be allowed to re-enter the GATT and should the GATT negotiation of the Uruguay Round be successful in significantly reducing trade barriers. In fact, China has been emphasizing grain self-sufficiency and has been able to meet domestic consumption in the last several decades without depending on large-scale importation of grain. Thus, the assumption of entire self-sufficiency may be relatively unimportant in the short run. In the long run, however, market reform and rapid industrialization may induce much more imports and export of food grains.

7.2 COMMENTS ON THE CHINESE AGRICULTURAL POLICIES

Despite the limitations discussed in the previous section, the models presented in the Appendix produce many sensible and consistent results. The main objective of the study is to test the effects of different policy scenarios on the patterns of grain production in China. If the ideal pattern based on a neoclassical economic norm is distorted, economic efficiency in the grain sector is reduced.

As discussed in Chapter A4, if the first model is treated as a basic one in which perfect competition is assumed, all the other models have, to a varying degree, net welfare losses to the society by introducing any government intervention. The production patterns of grain also shift with intervention.

The shifting of grain production pattern away from the perfectly competitive one described in model A2.1 (Chapter A2) due to intervention implies that individual regions cannot produce according to their comparative advantage, which explains the loss of economic efficiency. Thus, the estimated results from the models correspond to the two hypotheses set up in Chapter A1.

Since 1949, the government has used three major policy tools, i.e., consumer subsidy, production tax and production compulsory quota, to control grain production and marketing. Since consumer subsidy has been entirely under state control, it may not have any significant effect on the production pattern of grain because the state may be considered as a middleman between farmers and consumers or as a big trader in the market. However, this big trader can only buy a certain proportion of total output. Total consumption of the urban sector has also been restricted by grain rationing so that consumers with state subsidies have not been able to consume or purchase as much as they like at the subsidized grain price.

As noticed in Chapter A4, if the main objectives of the government are to maintain a high level of grain supply and a low price to urban consumers, each of the first three policy packages, i.e., production quota, consumer subsidy or the combination of these two policy tools (embodied respectively in models A2.2, A2.3 and A2.4 in Chapter A2), would be quite appropriate. All the three policy scenarios bring about greater grain production according to the quantitative analysis summarized in Table 7.1.

The total increase in grain supply ranges from 6.3 million to 14.9 million tons. Total estimated net welfare losses of these models to the economy are also moderate compared with those recorded in the last two policy scenarios embodied in models A2.5 and A2.6. Another common characteristic of the first three policy models is that consumers benefit significantly. Consumer surplus increases by as much as 20 billion yuan (model A2.4 – consumer subsidy and production quota).

However, with a production quota alone (model A2.2), the government is unable to squeeze any budgetary gain from the producers. The policy scenarios embodied in models A2.3 and A2.4 result in large budgetary costs to the state. Thus, the first three policy scenarios in models A2.2 to A2.4 directly contradict another objective of the

Table 7.1   Effects of policies on welfare and grain supplies (million yuan and million tons)

Effects on	M2	M3	M4	M5	M6
Producers	−8664	+2495	−362	−1883	−32967
Consumers	+8391	+17422	+20253	−35548	−6367
Budget	—	−20247	−20362	+35818	+38612
Net Loss	−273	−330	−471	−1613	−723
Supply	+6.3	+12.9	+14.9	−27.6	−4.7

Notes:
(a) M2–M6 are respectively models A2.2 to A2.6.
(b) Effects of different policies on producers, consumers, the state budget are the welfare gains or losses to the producers, consumers, the state budget and the society by comparing the corresponding models with model A2.1 – the basic model.
(c) All the values in terms of welfare and grain supply are national total welfare gains/losses and supplies obtained from models A2.2 to A2.6 as listed in Tables A4.11 to A4.15.
Sources: Tables A4.11 to A4.15.

government which is to extract resources from agriculture in order to support its ambitious industrialization programme. Thus, none of the first three policy packages has been ever adopted by the government since 1949. In fact, production quota has been implemented together with an indirect production tax (the policy scenario in model A2.6 in Chapter A2, to be examined later). Consumer subsidies have not been implemented in the manner described in models A2.3 and A2.4 in which grain market prices are brought down by a great margin for consumers through lowering regional production costs. Lower production costs are artificially created by imposing a unit product subsidy. In reality, the government never has implemented such a subsidy. However, the technical results of this manipulation are very similar to those produced by consumer subsidy.

Although consumers get most of the welfare gain in models A2.3 and A2.4, the benefits are equally spread to all groups of grain consumers, including the rural population. Since the grain subsidy programme is designed to benefit urban consumers alone, the policy scenarios in models A2.3 and A2.4 can hardly meet this objective, and this may be another reason why the government has not adopted these forms of intervention.

Although the first three policy scenarios have not been adopted by the government, the analysis of these policy options offers some very interesting and useful results. First of all, the computational results of these models can be used to demonstrate the opportunity costs of the other two policy options embodied in models A2.5 and A2.6 which have actually been adopted by the state. As will be noted below, a policy of heavy indirect taxation and its combination with a compulsory production quota usually incurs much greater net social welfare losses. It also depresses total grain production and farmers' incentive more than the first three policy scenarios (Table 7.1).

Secondly, models A2.2 to A2.4 quantify the costs and benefits of some important policy options which could be used for future intervention. Grain shortage during the four years 1985 to 1988 created many problems in the Chinese economy during that period. It triggered the historically high inflation in modern China and created a disastrous tension between the government and the people. This conflict between the people and government later on escalated into a large scale democratic movement led by the students of Beijing in 1989. After the Tiananmen incident, the government was forced to put more emphasis on increasing total grain output and production efficiency. Thus the policy options in models A2.2 to A2.4 have in fact partially reflected the policy changes adopted by the government since 1989: reinforcement of production quotas and lighter production taxation as opposed to heavy taxation with quota of the policy scenarios in models A2.5 and A2.6.

However, to understand the implications of the actual Chinese agricultural policies and make some comment on them, the following analysis will concentrate on the estimated results of models A2.5 and A2.6.

As discussed in the first part of the book, indirect agricultural taxation increases as the gap between the procurement price and market price widens. Therefore, the lower the procurement price, the more farmers will lose. Increasing losses in grain production inevitably discourage farmers' incentives, resulting in a loss of production and a loss of production efficiency. In our models, the loss of production efficiency is reflected in the distortion of production patterns. This is vividly demonstrated by the results from model A2.5. If we just focus on the effects on production and economic efficiency and ignore the welfare effects on producers, consumers and the budget, the policy scenario (only production taxation) in model A2.5 is the least desirable option among all policy scenarios as shown in Table 7.1.

The net loss to society from model A2.5 is more than 1.6 billion yuan, which is between two and six times as much as the losses of the other policy scenarios. The estimated loss of production is 27.6 million tons, which is more than 7 per cent of the total output in that year. Since grain self-sufficiency is one of the most important government objectives, the outcomes of this policy scenario are particularly unsatisfactory.

As for welfare redistribution, the budgetary gain is very significant, but the loss to consumers is almost equally large. If the loss to consumers were wholly compensated for by the government, both the government budget and consumers would not gain or lose under the policy scenario of model A2.5 as opposed to the basic model. Therefore, the total net loss to the society falls entirely upon the producers in terms of lower production efficiency. This is because the marginal lands in those high-cost regions become unprofitable for grain production as a result of taxation. These lands have to be sown with other crops which are less profitable than grain crops if grain production is not taxed but now become more profitable because grain crops are taxed. Thus producers suffer a double blow as a result of production taxation. They receive less than they should receive for their products; they are forced to produce less of the products (grain) in which they should have a productive advantage.

From the above analysis, we are able to draw our first comment on the Chinese grain policy:

Comment 1: Indirect agricultural taxation is one of the most important policy instruments used by the government to squeeze agriculture to support industrialization since independence in 1949. If the policy were implemented alone without any other instruments, there might not be any benefit or cost to consumers and the government budget if the loss to consumers were wholly compensated for by the gain to the budget. An indirect agricultural tax, however, could generate the most severe efficiency loss of grain production, and the net social welfare loss would be entirely borne by the producers. The significant loss in social welfare and severe reduction of grain output is mainly due to the diversion of agricultural land from grain crops to non-grain crops. This policy scenario does not help the government at all to raise (net) budgetary revenue from agriculture to support industrialization because the loss to consumers is almost equal to the gain to the budget. Its negative effects are obvious and severe in that grain output drops significantly and farmers suffer large income losses. Reduced grain production is directly contradictory to a broader national objective of food self-sufficiency.

From the above comment, it is obvious that the government would be very unwise to use indirect taxation alone without using any other policy instruments. But what other policy tools can be used by the government to supplement taxation? This depends largely on whether the government is in favour of the consumers or the producers. In China, the government has been apparently in favour of the consumers. The government is also very concerned with food self-sufficiency. Thus any policy instrument which can be used to ensure higher production and lower market price for the consumers will be the most ideal instrument. This turns out to be a compulsory production quota. Thus the policy scenario in model A2.6 which combines production taxation and quotas is found to be the most ideal policy option for the government. Indeed, this policy scenario is the actual policy option which has been used by the government for decades. The following analysis will be focused on the results of this policy option.

The computational results of model A2.6 help us to understand the working mechanism and the extent to which the combination of production taxation and quotas affects production, efficiency and the welfare of consumers, producers and the state budget. In the model, production taxation is incorporated in the form of lower producer price paid to farmers as this has been actually done by the government since the mid 1950s. Production quota is incorporated into the model in the form of a lower bound land constraint on grain crops. Thus whatever happens to the farm-gate price, farmers are forced to produce a minimum amount of grain in all regions (for a detailed description of the model, see Chapter A2).

Compared with the results of model A2.5, the net loss to the society and the reduction of grain output are greatly reduced in model A2.6 as seen from Table 7.1 (for a full theoretical analysis, see Section 4 of Chapter A1). The net loss to the society is reduced from 1.6 billion to 723 million yuan, and the loss of grain output is reduced from 27.6 to 4.7 million tons.

The budgetary gain only increases slightly but the loss to consumers is substantially reduced from 35.5 to only 6.4 billion yuan. If the loss to urban consumers were wholly compensated for by the government, the net gain to the budget would still be huge. The budgetary gain, however, is at the expense of grain producers. The loss to producers increases from 1.9 billion yuan in model A2.5 to 33 billion yuan in model A2.6. This explains how the Chinese government has been able to extract resources from agriculture without directly contradicting the broader objective of grain self-sufficiency.

Large extraction of resources from agriculture through the combination of production taxation and quota has enabled the government to subsidize urban and industrial development and achieve grain self-sufficiency. However, the policy imposes a tremendous burden upon the farmers. In addition, as production quotas are forced upon all the production regions, the policy as a whole induces great distortion to the production pattern across regions. As a result, potential economic efficiency through specialized production according to regional comparative advantage is lost.

So far, the discussion of the effects of this policy scenario has been focused on the year 1985 but the extent of distortion by the policy must be different as the magnitudes of taxation and quotas change over time. Higher taxation and greater quotas will induce more distortion whereas lower taxation and smaller quotas will do the opposite. Before 1979 indirect taxation was much higher and production quotas much greater and, coupled with the commune system, agricultural production and farmers' incomes were severely depressed. Between 1949 and 1978, more than 80 per cent of the farm land was sown to grain crops and the production of cash crops was greatly restricted (Chapter 1). Farmers' incomes remained almost unchanged in real terms at a very low level for three decades.

Empirical results observed during 1979 and 1984 show that the production pattern of grain has been improved so that land productivity and total grain output increased rapidly despite a steady decline in the area sown to grain crops. Improvement in land productivity and total grain output through more specialized production of farm crops was partly explained by the sudden increase in grain procurement price and some liberalization of self-sufficiency policy (quota) for every locality as mentioned in Chapter 1. Therefore, the computation results from model A2.6 are consistent with our argument and empirical observations. The relatively unsuccessful performance of agriculture and the stagnation of grain output during the 1985–88 period may be explained by the widening gap between the procurement and market prices of grain and the reduction of production quotas (Chapter 3). This argument is strongly supported by comparing the outcomes of model A2.5 and model A2.6 in Table 7.1.

The strong recovery of grain production from 1989 onwards also helps to explain the sensitive effects of tax policy. The changes made after 1989 were to reverse some of the undesirable effects caused by the changes in 1985. However, the principle of government policies has remained essentially the same as before, even after 1989. The underlying policy is a mix of indirect taxation and production quotas. Why the outcomes are distinctively different in different periods has been more or less explained by the magnitude of taxation. If the gap between procurement and market prices is reduced as experienced in 1979–84 and 1989–93, productivity and total output will be boosted. On the other hand, if the government allows the gap between the official and the market prices to widen as experienced in the pre-reform era and the period of 1985–88, the opposite effects are observed.

One may argue that our analysis on grain production has so far relied too much on these two policy instruments alone. It may be too naïve and narrow-minded to suggest that changing these two policies would solve all the problems in the grain sector. It is certainly true that there are many other factors to which can be attributed the fluctuations of grain production in China. However, we emphasize these two policy tools because they have been the predominant instruments which can be easily manipulated and have actually been used by the policy-makers to influence grain production and agricultural development. It does not mean that we should ignore other policy aspects which can potentially affect production and efficiency. Technical innovations and the adoption/diffusion of new technologies can also play a crucial role in the grain sector. Notwithstanding these aspects, our analysis will still have to focus on these two important policy tools.

The following comment summarizes the policy implications of this policy scenario:

Comment 2:   Compulsory production quotas have been used by the Chinese government to supplement indirect production taxation to reduce the negative effects on production and efficiency caused by the implementation of the latter instrument alone. There is no doubt that in theory or in practice, the combination of these two policy tools is much better than the single policy of indirect taxation in terms of efficiency, consumer welfare, production and tax revenue. However, producers have to bear greater losses. The analysis of this policy option supports our argument that the government has been sacrificing the farmers for rapid industrial growth. The impact on the pattern of grain production and farmers' incomes varies according to the degree of taxation and quota. The reduction of indirect taxation by increasing the procurement price in the early years of reforms (1979–84) induced grain production to become more specialized (and, therefore, more efficient) and resulted in fast growth of farmers' incomes. The widening gap between the state procurement and market prices after 1985 had led to a stagnation of grain production for four years from 1985 to 1988.

The reduction of a compulsory production quota in 1985 reinforced the loss of grain production and efficiency when the gap between the procurement and market prices was increasing. The reduction of a grain production quota enabled farmers to divert some grain fields to other crops whose marketing was entirely liberalized in 1985. Thus, the policy changes in 1985 (Chapter 1) might have enabled farmers to maintain sustained income growth, but the performance of the grain sector was very unsatisfactory. This argument was supported by the empirical results presented in Chapter 1 and by the computational outcomes from model A2.5.

The above comment is a mixture of theoretical study, empirical analysis and field observations. By referring to Table 7.1, it enables us to make the related comment given below:

Comment 3: There may be two alternative ways of increasing total grain output. One is to reduce indirect agricultural taxation by increasing the procurement price of grain and/or reduce the prices of inputs to raise farmers' incentives. The other is to impose greater production quotas. The first will lead to increasing the farmers' welfare and degree of specialization in production and greater efficiency. The second looks very similar to the old policy regime prevailing before the economic reforms. Greater quotas induce long run negative effects on specialization and efficiency. It also depresses farmers' income. Thus grain production would return to a vicious circle in which farmers are trapped in a 'poverty-cage', and the potential of economic efficiency through specialized production according to regional comparative advantage could not be wholly exploited. Reduced specialization would also contain the long-term growth of grain output although a compulsory quota may bring a sudden upsurge of total production in the short run.

After the policy changes in 1985 (see Section 4.2, Chapter 4), grain production fell sharply in 1985 and stayed below the 1984 record until

1989. During this period, many reports from the Chinese official media revealed that the problem of food shortage prevailed all over the country. Government officials of a number of regions, such as Hunan, Henan, Shangdong, Zhejiang, Jiangxi and Anhui urged the farmers to increase areas sown to grain crops in 1989. The central government was also forced to increase the procurement price of grain by 18 per cent in 1989 and stabilize the prices of inputs such as fertilizers. Credit at low interest rates was also made more easily available to grain producers. These changes of policy were a positive response of the government to strengthen the growth of grain production. Again, they involved two essential components: increase in acreage (production quota) and a reduction of indirect taxation.

Although an increase of 18 per cent in grain procurement price was much less than the inflation in 1988, which was 26 per cent according to official statistics, farmers responded positively to this price change. Grain production jumped from 399 million tons in 1988 to 414 million tons in 1989, the first time it had broken through the 1984 record of 407 million tons for six years. It surged to 446 million tons in 1990 and dropped back to about 435 million tons in 1991 (Table 2.11).

A strong recovery of grain production in the last few years is another great success of the reform policy. Adequate food supply ensured price stability and helped to sustain the speedy growth of the national economy. Total industrial output increased by more than 15 per cent per annum from 1989 to 1992. The rural industrial sub-sector grew by more than 20 per cent per year over the same period.[2] This achievement would not have been possible without a sustainable growth of agriculture in general and grain production in particular.

Notwithstanding the successful performance of grain production in recent years, many new problems are now emerging. First of all, the policy changes made in 1989 were again regarded as temporary methods to boost production. They were still restricted to the policy jargon described by the policy scenario in model A2.6. In other words, indirect taxation is still a strong element of the policy option. The new policy regime has only been trying to reduce the gap between the procurement and market prices instead of entirely eliminating indirect taxation. As long as indirect taxation is imposed, it has to be implemented with forceful production quotas. Thus grain production and marketing have not been entirely liberalized. This is an unresolved problem of the old policy regime. The only difference is that this problem could become much more disturbing and damaging to the economic progress of China under the new economic situation.

The economic strength of China is now far greater than a decade ago. The role of agriculture has correspondingly become less important in the national economy, but a recession in agriculture could bring about much more damage and destruction to the national economy. This is because Chinese consumers are accustomed to price stability and remain sensitive to price increases of basic consumer goods, especially foodgrains. The overheating of the industrial sector has induced high inflation in the last two years. The prices of many industrial items, such as construction materials and electricity, have increased by 15–25 per cent in 1991 and 1992. High inflation has not caused any slowdown of the economy in 1993. This is due to the incompetence and/or complacency of the central and local governments. Ample supplies and stable prices of grain and other food products due to a number of consecutive bumper harvests of grain crops have also helped to sustain the booming economy.

However, if an adequate growth (2–5 per cent per year) of grain output cannot be achieved in the coming years, and if the government is not able to control the investment thirst of firms and the local governments, food shortage is deemed inevitable. This will push up prices of both food products and industrial goods, resulting in another setback for the economy. According to past experiences, we have every reason to believe that after a few years of boom, China will suffer a major recession in agriculture followed by industry and trade.

The crucial question is whether this vicious economic circle can be avoided in China. Our answer is a firm *yes* provided the following two conditions are met: (a) agricultural producers have enough incentives to supply adequate grain and other agricultural products; and (b) the government is able to constrain investments and the hectic expansion of the industrial sector.

Unfortunately, the above two conditions are not easily satisfied in practice, due to a number of subjective and objective reasons. First of all the government lacks competence in managing the economy at the macro level, including effective control of credit, investment and project evaluations, etc. Secondly, both the central and local governments are often too ambitious and firms are usually too short-sighted. In the boom years, they tend to aim for unrealistic growth targets without paying attention to long-term profitability and sustainability, let alone the environmental damage.

For a country like China to become a real economic power will take many decades. Thus an annual growth rate of 6–10 per cent on a sustainable basis would be a great and realistic achievement. In practice, most policy-makers and local governments are always impatient to catch up with the most advanced economies. Thus a 10 per cent per year growth rate could seem too small to them. The problem is that in most years in China (even during the reform period), a high growth rate of the economy has not resulted in a comparable rise in the living standards of the Chinese people. Most of the growth has been accounted for by large-scale investments.

Investment and capital accumulation are essential for increased employment and the expansion of the economic base. The problem is that there are too many investment projects which cannot generate enough returns. Since the economic reforms, there has been a common phenomenon that many industrial enterprises have been able to expand their production capacities but have failed to increase or maintain their profitability. The government then has to bear an increasing financial loss made by the loss-making firms which are still expanding at an unbelievable speed. In 1992, as much as 20 per cent of the total government expenditure was used to cover the loss incurred by the state-owned industrial sector. The proportion of loss-making firms in the state sector has increased from about 30 per cent in the mid-1980s to two-thirds in 1992. Whilst the Chinese economy produced the best growth rate (over 12 per cent) in the world in 1992, the government was still running a huge budget deficit. Much of the deficit was required to meet the increased loss in these state enterprises and to finance over-committed investment projects.

Although the performance of the private and collective sectors (mainly rural township and village enterprises) is much more satisfactory, uncontrolled expansion of these sectors has caused large scale and rapidly increasing environmental damage. Their profit:capital ratio has also been declining over time, a strong indication of diminishing profitability of capital (see Table 3.6).

The thirst of investments in the industrial sector has a very damaging effect on agriculture, especially grain production. The desire to raise more monies for industrialization has led to the neglect and suppression of rural and agricultural investments. It also requires a consistent need to tax agriculture (mainly grain after 1985). Although grain production has been very satisfactory in recent years, there are signs that it may suffer another setback eventually if farmers' incentives are not increased. Due to the consecutive bumper harvests in the last few years, farmers in many areas have been given IOUs for their delivery to the state marketing agents. This partly reflects the many constraints we have identified in Chapter 4. It is also explained by the lack of cash held by many local banks and governments who are unable to pay the farmers instantly in cash because they have been committed to too many investment projects. IOUs reduce farmers' incentives and jeopardize future reproduction. The *People's Daily*, the official Chinese newspaper, reports that this problem has become so serious that the state council has issued a directive to order all the local authorities to pay the farmers promptly in cash for their deliveries. However, in China, 'an order is an order, practice is practice'. This means that local authorities can still do what they can do or will do.

From the policy point of view, it is woefully inadequate to encourage grain production through a number of unpredictable government decrees in such a huge country as China. It requires a consistent set of reform policies not against (or in favour of) farm producers. First of all, taxation has to be reduced. Secondly, large-scale public investments have to be made on agricultural infrastructure and marketing facilities. Thirdly, the central government has to have a long-term and consistent commitment to maintain the balance of agriculture and the rest of the economy. Grain, as a strategic product of the nation, can no longer be a major target for taxation. If necessary, it has to be supported (but not protected) with every possible means.

As mentioned in Chapter 4 the greatest constraint for the government to eliminate grain taxation is the need to subsidize urban consumers. We have argued that according to many theoretical and empirical studies, it is entirely possible and necessary to eliminate urban consumer subsidies. Since 1991, the government has been trying to liberalize grain marketing in a selected number of regions. By May 1993, it is reported that about 20 per cent of the original rationing coupons were eliminated.

The state employees were correspondingly compensated by wage increments. This is certainly a very important step towards liberalizing food marketing in China. Once the experiment proves successful, it will be extended to cover other areas. It is highly possible that in the foreseeable future, consumer food subsidies will be entirely eliminated.

The elimination of consumer subsidies will have the most significant implication for grain production and agricultural growth. There is no doubt that it will provide an essential foundation for the sustainable growth of grain production. However, how quickly consumer subsidies can be eliminated still remains a question. In addition, even if consumer subsidies are eliminated, there is still no guarantee that all the benefits can be passed on to the producers. Government can still tax agricultural and grain producers in other ways. Thus the real growth prospect of grain production remains to rely on the genuine commitment of government to support agriculture. In the long run, government support to agriculture and grain production will depend not only on price incentives but also on large investments in rural infrastructure, research, extensions, education, marketing facilities and so on.

7.3 PROPOSALS FOR FUTURE RESEARCH

The design of mathematical models presented in this study is largely constrained by data. As mentioned repeatedly, our models are partial because there is no adequate data to establish a multi-product and multi-regional spatial equilibrium model which would be more complete and realistic. Consequently, if we can find adequate data in the future, we have the following proposals to extend this research project towards a more 'ideal' target.

(1) Regional demand and supply functions should be re-estimated based on regional or, if possible, sub-regional data. Regional demand and supply functions for each individual grain crop and other crops should also be estimated before building a multi-regional and multi-crop model.

(2) More information about inter-regional flows of grain should be collected, including the means, the routes and the actual costs of transportation. The barriers to inter-regional marketing should be examined as well. The barriers may include local governments protecting regional interests without paying attention to the general goal of efficiency and specialization of the whole nation. This has been a common phenomenon in China and has been criticized by the central government and many economists who are more concerned with production specialization in agriculture.[3] Barriers may also be created by inadequate transport and other marketing facilities. Of course, the extent of constraint on inter-regional exchange may be different across regions.

Some regions, such as those situated in northeastern China, may have fewer restrictions imposed because they have a much better railway network. Other regions, such as those in the south and the west, may have more constraints imposed because the transportation system there is highly inadequate. We have not been able to consider this type of restriction in our models. Instead, our models assume free flows of commodity without any volume restriction at all. A more realistic model would take such a factor into account.

(3) Grain should ideally be disaggregated into individual products, such as rice, wheat, corn and others so that the substitution effects between different types of grain can be analysed by using a multi-crop spatial equilibrium model. Cash crops and other non-grain crops can also be incorporated to form a more complete model for the entire crop sector.

(4) The models presented in the Appendix could also be used to test the regional price policy of the government. For example, given the regional grain demand and supply functions, the optimal prices in different regions are obtained from the computation results. The regional price differentials can be used by the government to set the procurement prices for different regions. Since our major topic is not to deal with the policy of regional price differentials, we cannot have detailed analysis in this area. However, a relevant study on spatial equilibrium by A. J. Rayner proved that different policy options of regional price differentials had different impacts on the production pattern and economic efficiency of agricultural commodities such as milk (Rayner, 1974).

In a perfectly competitive market, regional price differentials are entirely determined by the transportation and marketing costs between regions. However, if prices are set by the government, the regional differentials may reflect some other policy objectives, such as regional subsidies or taxation. Consequently, the price system may result in a further distortion of the production pattern and efficiency losses. This is certainly another major problem of the Chinese agricultural policy (for a theoretical discussion, see the section on pan-territorial pricing in Chapter A1 of the Appendix).

(5) If inter-period storage costs of grain and the storage capacities in different regions are known, our 'static' one-period spatial equilibrium model can be enlarged to become a multi-period model which is 'dynamic' and can be used for future planning or forecasting. Such a multi-period spatial equilibrium model is particularly appropriate to test the medium and long term effects of policy changes on specialized production in agriculture or grain. However, the task of setting up such a model is a tremendous one which requires much more detailed and high-quality data concerning demand, supply, personal incomes, economic growth, population growth, storage facilities and costs, imports, exports and prices, etc., than is likely to become available in the immediate future.

7.4 FUTURE GRAIN POLICIES

Since the economic reforms, grain production in China has experienced two boom periods (1979–84, 1989–92) and one stagnant phase (1985–88). Of course, significant fluctuations of grain production are not unusual either in China or elsewhere in the world. However, in the analysis throughout this book, we have argued that much of the fluctuation of grain production in China since 1979 has been due to irrational policy changes.

Some policy changes reflect the conflict of government objectives. A typical example is the effective reduction of the grain procurement price in 1985 by the government to ease the budgetary burden incurred during the earlier years of reforms (1979–84). Other policy changes, however, reflected the incompetence of policy-makers in managing the economy. For example, it was undoubtedly true that the policy-makers were not aware of many unintended consequences of the policy changes in 1985 (see Chapter 4). In retrospect, the decision to cut the grain procurement price in 1985 was inappropriate but understandable because it had a clear objective to reduce government expenditures. The decision to reduce production quotas at a time when the gap between procurement and market prices was widening was certainly irrational because the immediate consequence was a double blow on grain output and economic efficiency.

Notwithstanding the many problems encountered during the reforms, agriculture and grain production in the last fifteen years has been highly progressive and successful. Grain output moved up a number of significant steps. In 1984, it broke the 400 million tons record. In 1990, it peaked at 446 million tons. These incredible records were achieved with a diminishing acreage allocated to grain crops. If further economic reforms can become more favourable to agriculture, it is highly likely that total grain output by the end of this century can reach the 500 million tons target set by the government without any acreage expansion. If it can be achieved without significant year-on-year fluctuations, China will be able to maintain a very high level of grain self-sufficiency.

In Chapter 4, we have argued that achieving higher levels of grain output is not only possible but also very beneficial to the national economy. From the efficiency point of view, China has a great advantage in rice production as reflected in the DRCs of rice. China also has slight advantage in wheat and other food crops as well. Comparative advantage has the most important implication for grain production in the long term. This is because future expansion of grain production will not be negatively affected by China's re-entry into the GATT. Thus the economic problem on grain production is not external competition. However, grain production has to face increasing internal competition for resources, especially land and labour.

Before the economic reforms, labour was abundant everywhere thus it was not a binding factor of grain production during the 1979–84 period. Due to the unprecedented expansion of rural non-farm activities, especially the rapid development of RTVEs, labour has become increasingly scarce. In fact, in some of the most advanced regions, grain production has completely lost its internal advantage (whether it has lost its external advantage is not clear and more difficult to justify). It has actually required significant subsidies from other production activities.[4]

The competition for land has been fierce and is increasingly so as industrialization and urbanization have acquired, and will continue to acquire, large areas of fertile land. The scale of decline in agricultural land is vividly demonstrated in Table 7.2. In 1991, some 302,200 hectares of arable land were destroyed for building factories and houses, accounting for 0.32 per cent of the total arable area in 1990. Although much of the area lost is compensated for by newly reclaimed land, there was still a net loss of 19,300 hectares in that year, or 0.02 per cent of the total arable area.

Table 7.2   Changes of arable land between 1990 and 1991 (1000 hectares)

(1)	Total arable area at the end of 1990:	95673.0
(a)	Total gain of area in 1991: **of which:** newly-reclaimed:	468.6 276.8
(b)	Total loss of area in 1991: **of which:**	488.1
	i) State construction projects:	71.7
	ii) Set aside for forestry:	129.7
	iii) Set aside for pasture:	56.2
	iv) Non-state construction projects:	230.5
(c)	Net loss of arable area in 1991: as % of total:	−19.5 −0.02
(2)	Total arable area at the end of 1991:	95653.5

Source: *ZGNYTJZL* (1991), p. 12.

The competition for land inside agriculture is also very severe. In many fast-growing regions, grain production has been losing its advantage to industrial and cash crops. Vegetables, fruit and fish (agricultural land can be dug to create fishing ponds) are much more profitable than grains in the outskirts of the urban and industrial areas. In areas distant to the urban centres, although the competition for land is less severe, farmers still have to decide how much land to allocate for industrial and cash crops in order to maximize their revenue. Thus the relative profitability of individual crops has a direct and crucial effect on land allocation. Grain production so far has been subject to compulsory quotas. The imposition of production quotas implies that grain production lacks comparative advantages as opposed to non-grain crops. Once the quotas are lifted, there will certainly be strong negative effect on grain production in China.

In short, although grain production in China has comparative advantage internationally, it lacks internal advantage. In the short and medium run, what will affect grain production most is internal factors instead of external competition as China is still outside the GATT. This raises a serious problem of how to sustain a satisfactory growth rate of grain output in the future. The key issue is how to direct future policies toward achieving this goal.

According to our analysis and past experience, we consider the following three points to be important and to need the most thorough and serious attention from the policy-makers.

Firstly, agricultural policy must be considered in conjunction with macro-economic policies. This point is rightly highlighted in a recent speech by Jiang Zheming, General Secretary of the Communist Party and Head of State. He said that to maintain fast and good development of the national economy, three issues must be emphasized this year (1993): (1) sincerely and absolutely to implement the policy methods suggested by the central Party committee to strengthen agriculture, trying to achieve a good agricultural harvest this year; (2) to strengthen macro-economic control, trying to realize a rough balance of aggregate demand and supply, and to achieve a more rational structure of the national economy; and (3) to deepen economic reforms (Jiang Zheming, 1993).

It is obvious that agriculture is still the most important concern of the government. The reason is simple. In China, there is still 80 per cent of the entire population making a living directly or indirectly related to agriculture. A good agricultural harvest is always crucial for the

Chinese economy. This does not require further explanation. What needs to be pointed out is that to achieve a good agricultural harvest, the latter two points of Jiang are critical. As mentioned in the previous section, uncontrolled industrial expansion not only requires a comparable growth of agricultural surplus, but also creates a tremendous constraint on agricultural development. This is why China has to strengthen macro-economic control and rationalize the structure of the national economy. This is exactly the second point of Jiang. As for Jiang's third point, methods of further economic reform remain very vague. However, as far as agriculture is concerned, it is necessary that more incentives should be given to farmers regarding product prices, supplies of inputs and marketing infrastructure. This implies that the government has to reduce agricultural taxation and divert more resources from the rest of the economy to agriculture. Farmers' incomes should be raised to narrow the income gap between agricultural and industrial workers gradually.

The second point of future grain policy is how to link grain production to other agricultural production activities. In other words, the incentive structure inside agriculture has to be readjusted carefully. Although market liberalization in 1985 abolished the state control on the production and marketing of all the non-grain crops, grain production has remained subject to quota restriction and taxation.

Production quotas and regional interests are two very important obstacles to raising the degree of production specialization among regions. Lack of specialization not only damages the prospect of diversification but also constrains further growth of aggregate grain output.

Unfair taxation against grain crops reduces the relative profitability of grain production. It is probably the most important disincentive to grain producers. Thus abolishing grain taxation should become the focus of future agricultural reforms.

The third point needing to be emphasized concerning future grain policy is that a set of laws are required to guard grain producers from being ill-treated by the local authorities and marketing agencies. It has been repeatedly reported that farmers are ill-treated by the local authorities and marketing agencies. Methods of unlawful treatment used against farmers have included numerous local taxes and duties, IOUs issued against grains delivered to the state, untimely and inadequate supplies of inputs caused purely by bureaucracy and corruption, unreasonable prices charged for inputs, etc. All these malpractices dampen farmers' incentives, reduce their incomes, and above all threaten sustainable growth of grain production.

Part III
Appendices: Empirical Results and Mathematical Models

Introduction

This appendix presents the geometric analysis of some important agricultural policies in China. Based on the theoretical framework, a few policy models are constructed to study the impacts of various policy scenarios upon production and resource allocation efficiency in the grain sector.

Chapter A1 provides detailed geometric analysis of a number of important policies and examines their impacts on production and efficiency. The main policy instruments under study include consumer subsidy, pan-territorial pricing, production quota, the combination of production quotas and indirect production taxation, input subsidy, and export taxation. These policy instruments have been extensively used by the government in various ways and different periods. Some of them, such as input subsidy and export taxation, have become less important in recent years due to market liberalization. However, most policy instruments still have direct and significant implications on grain production in China. The extent to which each individual instrument has affected grain production and resource allocation efficiency is discussed in detail.

The impact of each policy instrument includes the welfare effects on consumers, producers and the taxpayers. In terms of regional interests, the welfare effect of any instrument on a particular region depends on the net effect on the consumers and producers of that region with respect to the policy. As our main concern is about national interest, the detailed effects of policy instruments on individual regions are not discussed in this chapter although they appear clearly in the empirical analysis in the following chapters. The theoretical analysis in Chapter A1 reveals the common weakness of traditional policy analysis which usually ignores the variations of policy impact at the regional level.

Chapter A2 develops six policy models for the Chinese grain sector. These models aim to analyse the impacts of the most important policy instruments (production taxation, consumer subsidy and production quota) on grain production and efficiency in China. Each policy model consists of a theoretical framework which is then developed into a format suitable for direct empirical application. All the models, however, have to be simplified in one way or another owing to data limitation. The usefulness and disadvantages of the models are discussed in detail.

Chapter A3 is devoted to processing the data used in the model simulation. For simplicity, the country is demarcated into 26 demand and supply regions. Demand functions are estimated for individual regions under the assumption that different regions face the same demand coefficients (linear) of price and income. In other words, the response of grain demand to incremental price and income is the same for all regions. This is so done because of the lack of time series data which prohibited us from estimating regional demand functions individually. As we could not make the same assumption in estimating regional supply functions, we have to make an even stronger assumption that regional supply function is entirely horizontal. The level of regional supply is dictated by the cost of production. The complementary assumption on regional supply is that each individual region can only produce a maximum fixed amount of grain output. This is imposed by setting land upper bound constraints. The data used are the cross-sectional data for 1985 which is regarded as the most important year in terms of policy changes in the reform period. The year 1985 witnessed several policy changes which had positive intentions, but yielded poor results. The analysis of this appendix is exactly aimed at identifying the weakness of those policy changes and summing up the lesson to be learned by the policy makers.

The models are solved by a computer package called LINDO which is designed to solve both linear and quadratic optimization problems. The computational results for each model are reported in Chapter A4. The effects of each policy scenario include the changes of production and the changes of consumer/producer surplus across regions. Following the conventional approach, the basic model used for comparison is that of no intervention. Comparisons are then made between the results of policy models with intervention and those of the basic model without intervention. Thus the costs and benefits of each policy model become quantitatively clear.

A1 Geometric Analysis of Selected Policies

Below we examine the effects of different policy tools on grain production and the welfare of producers, consumers, and taxpayers at the national level.

The policy instruments to be discussed in the following sections include: (1) consumer subsidies; (2) pan-territorial pricing; (3) production quotas; (4) the combination of indirect taxation and production quotas; (5) input subsidy; and (6) export taxation. As these policy instruments are discussed separately, the analyses will be partial in that each policy instrument only reflects one single aspect of the actual market system where a number of instruments interact.

A1.1 CONSUMER SUBSIDY

A1.1.1 A simple framework

Consumer subsidy in this section is defined as a policy instrument employed by the government to support the living conditions of a particular group of people in the society or the whole population through ensuring lower than free market prices of goods sold to consumers.

There are a number of objectives associated with consumer subsidy:

(a) to maintain a certain living condition of the consumers;
(b) to ensure price stability in the market place and reduce the risk of hunger and malnutrition;
(c) to alleviate inequality and favour the disadvantaged groups of the population.

In most developing countries, consumer subsidy involves the provision of cheap food and other agricultural commodities by the government to consumers. The following analysis assumes that consumers pay a lower than market price when they buy their food from the state or state marketing agencies without any other policy restrictions such as rationing. Of course there are a number of other policy instruments, apart from consumer price regulation, which can be used by the government to implement the consumer subsidy programme. Consumption rationing, for example, can be used to reduce the total amount of subsidy and the pressure on food demand. Compulsory procurement from the producers can help reduce the budgetary cost of consumer subsidy by taxing the producers: that is, the producers can be forced to bear part of the subsidy costs.

The consequences of consumer subsidy can be generalized as follows.

(a) Consumer surplus will increase and the amount of welfare gain to consumers depends on the quantity consumed and the difference between the official and free market prices. In the following analysis, the free market price refers to the border parity price for consumers.

(b) The producers may or may not lose, depending on whether the government introduces other policy instruments such as compulsory procurement to support the consumer subsidy programme. In the simplest case, it is assumed that producers receive the parity price for their produce so that they will not lose.

(c) Budgetary cost increases and it is borne by the taxpayers.

(d) Low consumer price induces excessive domestic demand, resulting in more deficit or less export of those commodities under concern.

For simplicity, it is assumed that the country is a net importer of food and subsidy reduces the consumer price from the import parity price, P_w, to a lower level, P_s, as illustrated in Figure A1.1.

Producers still receive the import parity price for their product whether delivered to the state or sold to the market. Thus the government has to bear the entire cost of food subsidy. The country under study is 'small' and faces a perfectly elastic supply curve for its import at P_w.

(1) Supply, demand and import without food subsidy

Figure A1.1 illustrates the supply and demand relationship of a deficit country. D and S represent the demand and supply curves respectively. At P_w, domestic demand and supply are respectively Q_d and Q_s without consumer subsidy. The deficit between domestic demand and supply $(Q_d - Q_s)$ has to be imported from the international market.

Figure A1.1   A simple working mechanism of food subsidies

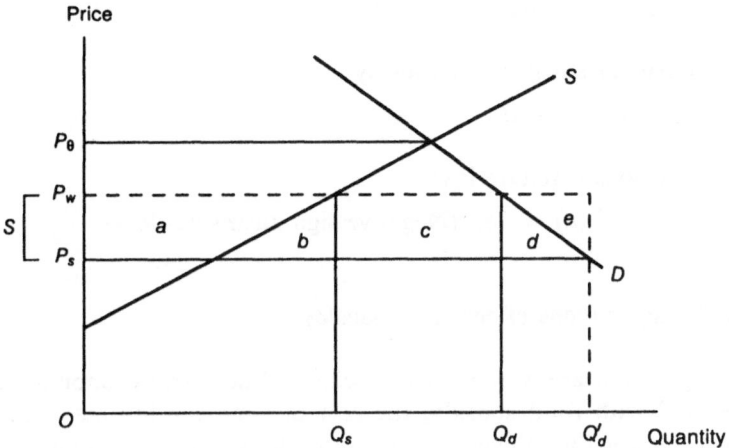

(2) The effects of consumer subsidy on domestic demand and import of food

Suppose domestic consumers buy food at the price P_s which is lower than the import parity price P_w by $S = (P_w - P_s)$ but producers still receive the import parity price P_w for their product, domestic consumption will increase from Q_d to Q'_d and domestic supply will remain unchanged at Q_s. The deficit of food in the domestic market will correspondingly increase from $(Q_d - Q_s)$ to $(Q'_d - Q_s)$.

(3) The effects of consumer subsidy on the welfare of domestic producers, consumers and tax payers (the state budget)

Since domestic supply remains unchanged with food subsidy, the welfare effect on producers remains unchanged. The entire cost of subsidy has to be borne by the budget or taxpayers.

The total subsidy cost is the sum of the areas *a*, *b*, *c*, *d* and *e*. Consumer surplus increases by the sum of *a*, *b*, *c* and *d*. The net social welfare loss is area *e*. It arises because the total gain to consumers is less than the total subsidy cost. The calculations of costs and benefits are demonstrated as follows.

(a)   Subsidy cost

$$a + b + c + d + e$$

(b)   Consumer surplus increases by

$$a + b + c + d$$

(c)   Total social welfare loss

$(b) - (a) = -e$ (Negative sign means net loss)

A1.1.2   Implications of consumer subsidy

Although there are a number of social, political and economic goals associated with food subsidy, the cost of such subsidy may become unbearable for the budget. Food subsidy may also create many other undesirable consequences in the process of economic development, which are discussed below:

(1)   High cost to the budget

In China, consumer subsidies consisted of 20–30 per cent of the industrial wage income throughout the 1980s and almost 18 per cent of the state budget (Chapter 4, Table 4.1; and Yao and Colman, 1990). This sizeable subsidy is one of the major factors contributing to the chronic problem of budgetary deficit. Similar problems have been encountered by many other developing countries.[1]

(2)   Implementation problems when food rationing is used to restrict the cost of subsidy

A critical problem of food rationing is the difficulty in identifying the target groups. For example, the lack of base-line data in the less developed countries (LDCs) on household income and consumption patterns makes it extremely difficult to delineate those sections of the population most vulnerable or at risk from changes in policy direction. The cost of distribution may become so high that the government finds it uncontrollable. As a second-best response governments apply a universal rationing pattern to the whole segment of the population such as the 'urban household' without taking differences in income levels into account. This can easily undermine the initial equity objective of the food subsidy programme. In some cases, food subsidy may well worsen income distribution, as in the case of China where food subsidies are exclusively provided for the urban dwellers who are often better off than their rural counterparts.

(3) Decline in food self-sufficiency

Subsidy encourages excessive consumption and if it is complicated by compulsory procurement from the domestic producers it may result in depressed domestic production, leading to an increasing gap between domestic supply and demand. More imports will be required to fill such a gap.

(4) Once implemented, food subsidies are very difficult to withdraw

Many countries have experienced extreme difficulty in retreating from the commitment of food subsidies which have been regarded as their 'God-given right' by consumers. Food subsidy has been a sensitive political and economic problem in the experience of most LDCs. In political or social terms, food subsidy has been used by some countries as a means to ensure food security for the vulnerable groups and its sudden withdrawal will violate the initial objective and incur resistance from people who have been used to the old system. In economic terms, many countries have used food subsidy as a means to ensure low industrial wages and its abolition requires readjustment of the existing wage structure which may induce inflationary pressure on the economy. This is one of the difficult problems in China during economic reforms.

A1.2   PAN-TERRITORIAL PRICING (PTP)

A1.2.1 Introduction

Pan-territorial pricing: A uniform price is imposed for the producers (product price) or the consumers (consumer price) for all the localities within the country regardless of the differences in regional demand and supply equilibria and inter-regional transportation costs.

In a free market system with perfect competition, the local market prices in different regions are determined by the local demand and supply in the absence of inter-regional trade. With inter-regional trade, producers in the surplus regions will ship their products to the deficit areas if the prices in these areas are higher than those in the surplus regions by at least the unit transportation and handling cost. An ideal country-wide equilibrium will be reached only if there exists a spatial pattern where the price differential between any pair of regions is not more than the unit transportation and handling cost between them.

Social and political arguments for PTP are based on regional equity and balanced regional development. In most poor countries, because of inadequate transportation and the long distances between the production areas and the urban centres, surplus production in the remote regions may be highly unattractive if there is no government support. A primary objective of PTP is thus to stimulate production in these remote areas.

Economic argument in favour of PTP arises from the fact that most countries implementing such a policy have done so to achieve food self-sufficiency.

One obvious consequence of PTP is that the government may have to bear the inter-regional transportation and handling costs.

Production in regions close to the consumption centres is generally suppressed while production in the remote regions is encouraged.

Consumption in the surplus regions (remote regions) is generally discouraged while in the deficit regions it is encouraged if the cost of transportation and handling is wholly subsidized by the government.

A1.2.2 The working mechanism

To understand the working mechanism of PTP, it is necessary to discuss the economics of spatial equilibrium and analyse the costs and benefits of interregional trade.

(1) The economics of spatial equilibrium

Suppose the country under study is divided into two regions with one being food surplus and the other food deficit, as is illustrated in Figure A1.2.

D_1 and S_1 are respectively the demand and supply curves in region A, D_2 and S_2 are respectively the demand and supply curves in region B. To understand the economics of spatial equilibrium, Figure A1.2 has to be analysed according to the following steps:

(a) Local prices, demands and supplies without inter-regional trade

If there is no trade between regions A and B, the local equilibrium prices for regions A and B are P_1 and P_2 respectively. The demand

and supply in region A are both equal to Q_1 and the demand and supply in region B are both equal to Q_2.

(b) Regional equilibrium prices, demands and supplies with interregional trade

Case 1: Trade without transportation costs

Suppose there is no transportation cost between regions A and B, producers in region A will find it profitable to ship part of their product to sell in region B where the local market price is higher than that in region A. To maximize their profit, producers in region A will continue shipping their product out of the region until both regions have exactly the same price. In Figure A1.2, the optimal equilibrium price for both regions is P_0.

The amount of export from region A is $(S_1 - D_1)$. This is equal to the amount of import by region B $(D_2 - S_2)$. S_1 and D_1 are respectively the optimal regional supply and demand for region A. S_2 and D_2 are respectively the optimal regional supply and demand for region B.

Figure A1.2   Effects of pan-territorial pricing

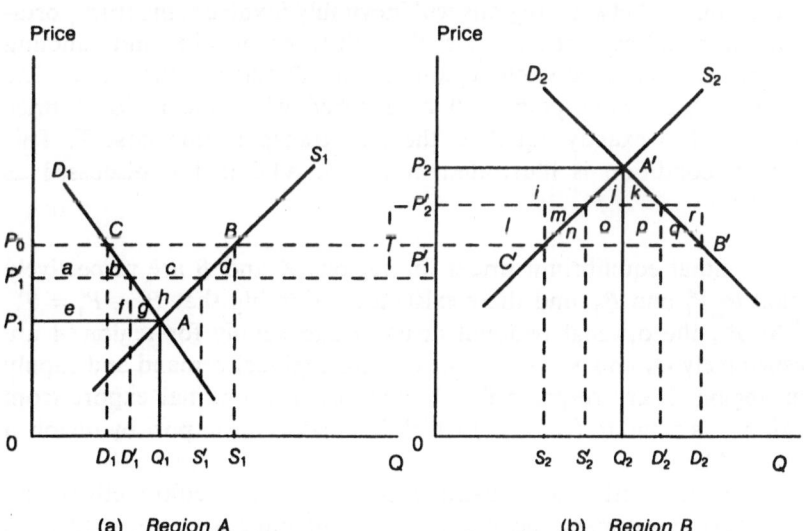

(a)   *Region A* (b)   *Region B*

Compared with the non-trade scenario, the welfare effects on producers, consumers and taxpayers can be analysed as follows:

For region A, producer surplus increases by $(a+b+c+e+f+g+h)$. Consumer surplus decreases by $(a+b+e+f+g)$. Net social welfare gain is $(c+h=ABC)$.

For region B, producer surplus decreases by $(i+l+m)$. Consumer surplus increases by $(i+j+k+l+m+n+o+p+q)$. Net social welfare gain increases by $(j+k+n+o+p+q=A'B'C')$. Total net social welfare gain is equal to the sum of net social welfare gains of both regions A and B. That is: $(c+h)+(j+k+n+o+p+q) = ABC + A'B'C'$.

Both regions have net social welfare gains under the scenario of inter-regional trade. Region A gains because the gain to producer surplus is greater than the loss to consumer surplus. Region B gains because the gain to consumer surplus is greater than the loss to producer surplus.

Case 2:   Trade with transportation costs

Case 1 can only be used for a hypothetical analysis. In reality, any transshipment between regions will inevitably involve some transportation and handling costs. Suppose the unit transportation and handling cost is a constant between regions A and B and is equal to T. The optimal equilibrium point will be reached when the regional price differential is exactly equal to the unit transportation cost T. This optimal condition is illustrated in Figure A1.2 and is discussed as follows:

The optimal equilibrium prices for regions A and B are respectively equal to P_1' and P_2', and there exists a relationship that $P_2' - P_1' = T$.
  At P_1', the optimal regional demand and supply for region A are respectively D_1' and S_1'. At P_2', the optimal regional demand and supply for region B are respectively D_2' and S_2'. The optimal export from region A is equal to $(S_1' - D_1')$ which is equal to the import by region B $(D_2' - S_2')$.
  Compared with the non-trade scenario, the welfare effects on producers, consumers and the tax-payers of trade with transportation costs are analysed as follows:

For region A, producer surplus increases by $(e + f + g + h)$. Consumer surplus decreases by $(e + f + g)$. The net social welfare gain increases by h. For region B, producer surplus decreases by i. Consumer surplus increases by $(i + j + k)$. The net social welfare gain increases by $(j + k)$. The total net social welfare gain is equal to the sum of the net social welfare gains to both regions A and B: that is $(h + j + k)$.

According to the neoclassical economic theory, case 2 is considered to be the most efficient spatial equilibrium pattern under free trade and perfect competition. On the one hand, inter-regional flows of commodity are entirely free.[2] On the other, the quantity of inter-regional commodity flows is limited by the existence of transportation costs. The combination of these two conditions leads to two basic conclusions concerning spatial equilibrium:

(a) Producers in low-cost production areas tend to ship their product to the high-cost production areas but inter-regional trade can only occur if the price of the deficit (high cost) region is high enough to cover the transportation and handling cost and the local price of the export region.

(b) At equilibrium, if there is any trade between a particular pair of regions, the price differential between these two regions is exactly equal to the unit transportation and handling cost between them. If there is no trade between a particular pair of regions, the price differential between these two regions is either less than or just equal to the unit transportation and handling cost between them.

(2) PTP within the context of spatial equilibrium

The principle of PTP is that every region faces the same price regardless of local demand and supply equilibria and inter-regional transportation costs. The simplest analysis on PTP can be made on the basis of assumptions implied in Figure A1.2 listed below:

(a) the inter-regional transportation and handling costs are entirely borne by the government budget or by the tax payers;

(b) producers in the low-cost production (surplus) regions ship their product to the high-cost (deficit) regions until such an equilibrium point is reached at which all regional prices are exactly equal.

In Figure A1.2, the ultimate equilibrium price for both regions A and B is equal to P_0. The total amount of export from region A is $(S_1 - D_1)$ which is equal to the total amount of import by region B $(D_2 - S_2)$.

Compared with case 2 in A1.2.2, the welfare effects on producers, consumers and taxpayers in each region can be discussed as follows:

For region A, producer surplus will increase by $(a + b + c)$. Consumer surplus will decrease by $(a + b)$. The transportation and handling subsidy is $(b + c + d)$. The net social welfare loss is equal to the transportation and handling subsidy plus the consumer surplus loss minus the producer surplus gain: $(a + b) + (b + c + d) - (a + b + c) = b + d$. The loss of b and the loss of d are respectively due to the excessive supply and suppressed demand in region A. For region B, producer surplus decreases by $(l + m)$. Consumer surplus increases by $(l + m + n + o + p + q)$. The transportation and handling subsidy is $(m + n + o + p + q + r)$. The net social welfare loss is equal to the loss of producer surplus plus the transportation and handling subsidy minus the gain to the consumers: $(l + m) + (m + n + o + p + q + r) - (l + m + n + o + p + q) = m + r$. The loss of m and the loss of r are respectively due to depressed supply and excessive demand in region B. The total net social welfare loss is equal to the sum of the net social welfare losses to both regions A and B, or $(b + d + m + r)$.

(3)   Implications of PTP

The analysis in section A1.2.2 can be generalized for a multi-region case study if we want to quantify the total social cost of PTP. The policy of PTP triggers a number of social and economic issues.

(a)   It affects the production pattern of crops. More production will be concentrated in the remote or the low-cost production areas at the expense of the high-cost areas or those closer to the consumption centres.

(b)   There will be a distinct redistribution effect on the producers and the consumers in different regions. For those regions which produce more and consume less (low cost or remote regions), the producers will gain while consumers lose. For those regions which import more and produce less, consumers will gain while producers lose.

(c) The cost of transportation and handling subsidy is always greater than the net gain accruing to regional producers or consumers. Thus, there is always a net social welfare loss which should be used as a trade-off to justify the non-economic goals such as regional balanced development or regional equity intended to be obtained by the PTP policy.

The cost of transportation and handling subsidy may be particularly high for most agricultural commodities, especially food crops, which are of low value and high bulk. PTP policy tends to encourage food crop production in the remote regions but the cost of transportation subsidy may become unbearable. Efforts to reduce such subsidies will discourage food production in those regions which already have benefited from the policy. It will be difficult for producers to change their production pattern within a short period of time. Farmers, smallholders in particular, will suffer immensely from such an adjustment programme unless a complementary method is employed to help them to diversify their product mix, e.g. encouraging the production of non-food crops which have a higher value:volume ratio and low component of transportation cost (for the Chinese experience, see Chapter 4, Section 4.2.4).

A1.3 COMPULSORY PRODUCTION QUOTA

Compulsory production quota is a special means by which governments force producers to produce a certain amount of a particular product. The product can be a grain or a cash crop.

In China, there are two reasons for imposing compulsory production quotas. One is to increase food self-sufficiency in all the localities. The other is to ensure that the government is able to secure a certain amount of product from all regions. In the reform period, restriction on grain production has been slightly relaxed. However, as the state still exercises mandatory procurement, a compulsory production quota is still effectively in place.

The effect of a compulsory production quota can be indicated by the following theorem:

Theorem A1.1:   In the short-run situation, a compulsory production quota imposed on every region generally increases total grain supply but it distorts the production pattern so that high-cost production regions are forced to increase grain production at the expense of cash crop production, while the low-cost regions may be forced to reduce their grain supply and shift resources to the production of cash crops.

This theorem is easy to understand but it may be difficult to prove it through a two-dimensional diagram. Figure A1.3 illustrates the case for a high-cost production region where a naive (but logical) assumption is imposed that the change in grain supply of this region does not have any effect on regional and national equilibrium prices of grain before and after the introduction of a compulsory production quota, i.e., a particular region in this figure is a 'very small' region in terms of grain supply compared with the whole nation.

Figure A1.3   Compulsory quota and a high-cost (deficit) region

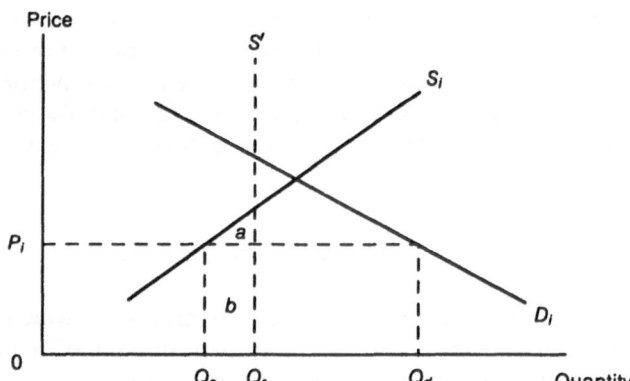

Suppose S_i and D_i are respectively the supply and demand curves of region I which is a high-cost and deficit region. At equilibrium with perfect competition, its equilibrium price is determined at P_i. The total demand in this region is Q_d and the total supply is Q_s. The differential between Q_d and Q_s is met by the shipments from surplus regions, assuming the country is a closed economy or it is food self-sufficient.

If a compulsory production quota requires that region I produce at least Q_r and, according to the above assumption, total demand and regional equilibrium price remain unchanged at Q_d and P_i, the cost and benefit for that region with a quota can be given as follows:

(a) Resource cost is equal to $(a + b)$. This may be considered to be the opportunity cost of cutting the production of the non-grain crop in that region to increase grain supply.

(b) Benefit is equal to b. This is the extra income of grain producers in that region.

(c) The net loss to the region or producers is the triangular area a.

In fact, if all the high-cost and deficit regions produce more under the quota, total supply of the whole nation will increase. This will inevitably lead to an overall depression of the market equilibrium prices. Thus, the cost and benefit analysis will be much more complicated than what is suggested in Figure A1.3.

A1.4 PRODUCTION QUOTA AND INDIRECT TAXATION

A production quota is defined in section A1.3. That is, every region has to produce a minimum amount of output by mandatory measures. Indirect production taxation means that producers are paid a lower than market price for the product delivered to the state.

The consequences of a compulsory production quota are discussed in Section A1.3. The consequences of an indirect tax, if it is imposed alone, would be as follows:

- Total domestic production would be reduced.
- The country will have less to export if it is an exporting country, or will have to import more if it is an importing country.
- Consumers may not lose or gain if the country maintains the same price through border trade to accommodate the shortfall created by reduced domestic production.
- Producers would lose, due to lower prices and reduced output.

In a country where food self-sufficiency is a top priority, a shortfall of domestic grain production is highly undesirable. To reduce this undesirable effect, a compulsory production quota has to be imposed. This compulsory production quota can be regarded as a supplementary or secondary instrument to sustain indirect taxation without sacrificing domestic production. In addition, the imposition of a production quota also helps to reduce the net social loss induced by a high indirect production tax alone.

The effects of the combination of these two instruments can be summarized by the following theorem:

Theorem A1.2: Under the circumstances of a high indirect production tax by paying a much lower than market price to producers, total grain supply and social welfare losses are substantial unless a compulsory production quota is also implemented. In other words, a compulsory production quota is able to reduce the negative effects on grain supply and social welfare induced by indirect production taxation.

This theorem can be proven by employing a simple framework illustrated in Figure A1.4. S and D represent respectively the aggregated national supply and demand curves. P_m and Q_m represent the market equilibrium price and quantity of grain traded under the assumption of a 'closed' or 'self-sufficient' economy. If the government pays the producers at price P_r, it imposes an indirect tax which is equal to $(P_m - P_r)$. With this simple framework, the following two cases examine the costs and benefits of two policy methods: i.e., indirect taxation alone and the combination of indirect taxation with a compulsory production quota.

Case 1:   Consequences of indirect production taxation are as follows:

(a)   total grain supply is reduced from Q_m to Q_r;

Figure A1.4   Effects of indirect taxation and compulsory quota on grain supply and welfare

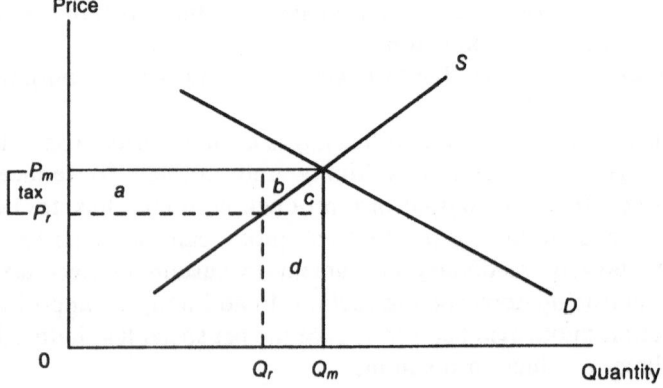

(b) the government has a tax income $= a$;
(c) producer's surplus loss $= a + b$, assuming the resource cost $(c + d)$ can be exactly recovered from other farming activities;
(d) if the government imports an equivalent amount of grain $(Q_m - Q_r)$ at price P_m to compensate the contraction in domestic supply, total foreign exchange expenditure is $(b + c + d)$ which can be wholly recovered if the government sells the imported grain at the same price P_m;
(e) the net social welfare loss $=$ producer's loss $-$ budgetary gain, or $= (a + b) - a = b$.

Case 2:   If the government can force the producer to produce Q_m and sell their product to the state at price P_r, then the consequences are as follows:

(a) total grain supply remains unchanged at Q_m;
(b) producer's loss $=$ a + b + c;
(c) budgetary gain $=$ a + b + c;
(d) total net loss to the society is zero.

Compared with Case 1, Case 2 has the following advantages:

(a) there is no reduction (or at least less reduction) in total domestic production;
(b) there is no net social welfare loss (or at least less net loss) if the handling cost of the policy tools is ignored;
(c) the government has a much greater budgetary gain.

The disadvantage in Case 2 is that producers' loss increases. This increased producers' loss may induce overall depression on agriculture and grain production as a whole in the long term. Thus, the long-run consequences will be more complicated. Such long-run consequences and their policy implications will be discussed in detail in the following chapters.

A1.5   INPUT SUBSIDY

A1.5.1   Introduction

Input subsidy is one of the most commonly used policy instruments to assist producers to increase their income and production. Agricultural inputs include fertilizers, insecticides and other chemicals, credit, irrigation, land, machinery, etc. Before 1985, input subsidies in China were used by the government to partially compensate the losses to producers through indirect production taxation. On the one hand, the government imposed heavy indirect taxation on grain producers; on the other, it sold various inputs, especially chemical fertilizers and insecticides, to farmers at fixed and low prices. The exact amount of subsidized inputs supplied to producers was conditional on the mandatory quota delivered to the state. Thus, input subsidy was an effective and positive policy instrument to ensure the success of compulsory procurement and indirect production taxation.

Policy changes in 1985 effectively abolished subsidies for most inputs in most parts of the country. As a result, input prices have increased steadily since 1985 (Chapter 4). Therefore, input subsidy in China is not an important policy instrument any more. However, it is still one of the most important policy tools in many developing countries. In addition, because input subsidy has a number of advantages as will be discussed below, it may still be a useful method for the Chinese government to assist farmers in the remote or backward areas.[3]

The analysis in this section is based on the following assumptions:

(a)   farmers buy fertilizers at a subsidized price to produce food crops;

(b)   the country under study is a food deficit country and its food production does not affect the price in the world market: that is, the country is facing a constant international price.

With these two assumptions, the consequences of food subsidy can be summarized as:

(a)   an increase in farm profit;

(b)   an increase in the quantity demanded of the cheaper (in this case, the subsidized) input;

(c)   an increase in the quantity supplied and improvement of food self-sufficiency;

(d)   either an increase or a decrease in the quantity of other inputs, depending on whether the tendency for more of all inputs to be used because of the higher level of output outweighs the tendency for the cheaper input to be substituted for all other inputs.

A1.5.2 Fertilizer subsidy

From a producer's point of view, fertilizer subsidy makes production costs lower and the normal supply curve is shifted to the right. This is illustrated in Figure A1.5. The impacts of input subsidy on domestic supply, demand and social welfare are analysed in detail below.

Figure A1.5 The working mechanism of input subsidies

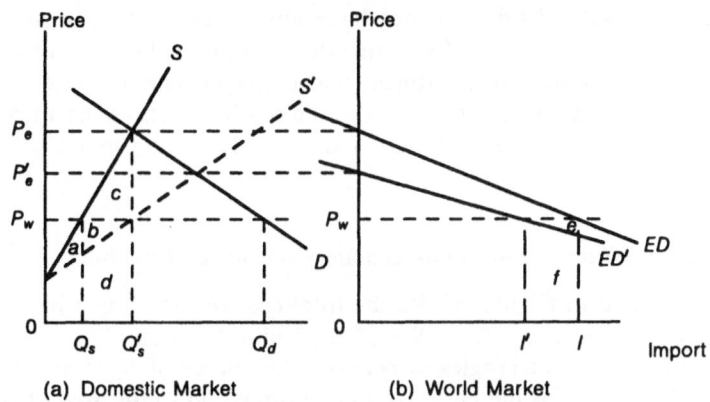

(a) Domestic Market   (b) World Market

(1) Effect of input subsidy on production, demand and trade

Case 1: Domestic supply, demand and import without input subsidy

In the domestic market (Figure A1.5a), suppose the supply and demand curves of food without subsidy are represented by S and D respectively. The domestic equilibrium point is reached at price P_e which is higher than the world market price P_w. If free competition prevails, the country under study will have to fill the deficit between domestic demand and supply based on P_w. The exact amount to be imported is $(Q_d - Q_s)$. Q_d is domestic demand and Q_s domestic supply at P_w.

The amount of food import without subsidy is equal to OI as shown in Figure A1.5b in which ED represents the excess demand curve at P_e.

Case 2: Domestic supply, demand and import with input subsidy

With input subsidy, because production cost is reduced, the supply curve is shifted from S to S' (Figure A1.5a) and the new equilibrium price at the domestic market decreases from P_e to P'_e. At price P'_e the excess demand curve in Figure A1.5b shifts from ED to ED'. Domestic supply at the world price increases from Q_s to Q'_s but domestic demand remains unchanged at Q_d. As a result, the amount of food import is reduced from $(Q_d - Q_s)$ to $(Q_d - Q'_s)$, or from OI to OI .

(2)   Welfare effect of input subsidy

Because the world price is used as the market clearing price in this analysis, input subsidy will not have any welfare effect on domestic consumers. The analysis of welfare effect of input subsidy can therefore be focused on domestic producers and taxpayers (or the state budget). However, when the analysis shifts to the border, there exists a trade-off between domestic costs (in terms of subsidy) and foreign exchange savings.

(i)   Effect of input subsidy on producer surplus and the budget

As illustrated in Figure A1.5a, the total cost to the budget in order to bring the supply curve from S to S' is equal to the area of $(a + b + c)$. a, b, and c are three triangles in between the curves of S, S' and Q'_s.

Producer surplus increases by an amount equal to the area of $(a + b)$ because the price received by the domestic producers is still P_w with input subsidy. The area c of budgetary cost is not recoverable and becomes a deadweight loss. The cost and benefit of input subsidy can be summarized as follows:

(a)   Total cost of subsidy

   $-(a + b + c)$

(b)   Increase of producer surplus

   $+(a + b)$

(c)   Deadweight loss or net social welfare loss

   $-c$

(ii) Effect of input subsidy on domestic cost and foreign exchange saving

As discussed in Chapter 6, input subsidy raises domestic production and reduces the amount of food import. As a result, foreign exchange outlay for food import is curtailed. The amount of foreign exchange saving due to input subsidy can be illustrated in Figure A1.5b as the area of $(f + e)$ which is the product of the quantity of reduced import induced by input subsidy multiplied by the world price P_w, i.e., $(f + e) = Pw^*(I - I')$.

The total domestic cost required to reduce the amount of import from I to I' is equal to the area of $(d + b + c)$. Because $(I - I')$ is equal to $(Q'_s - Q_s)$, $(f + e)$ is equal to $(d + b)$. Therefore, the domestic costs of c is not recoverable from foreign exchange saving. The cost and benefit analysis on trade due to input subsidy can be summarized as follows:

(a)   Total domestic cost of input subsidy to reduce import

$$-(d + b + c)$$

(b)   Total foreign exchange saving of reduced import due to input subsidy

$$+(f + e) = (d + b)$$

(c) Total net social welfare loss

$$-c$$

A1.5.3   A trade-off function of input subsidy

From the welfare analysis in sub-section A1.5.2, there are a number of welfare trade-offs arising from input subsidy:[4]

(a)   the trade-off between increased producer surplus and budgetary cost;
(b)   the trade-off between increased domestic cost and foreign exchange saving;
(c)   the trade-off between budgetary cost and foreign exchange saving.

Figure A1.6 demonstrates the trade-offs between the budgetary cost (BC) and foreign exchange saving (FE) with different amount of input subsidies. In Figure A1.6, input subsidy increases from (I) to (IV) and domestic production increases from Q_{s0} to Q_{s4}. As the subsidy (budgetary cost) increases, foreign exchange outlay decreases. This

Figure A1.6 Changing levels of input subsidies

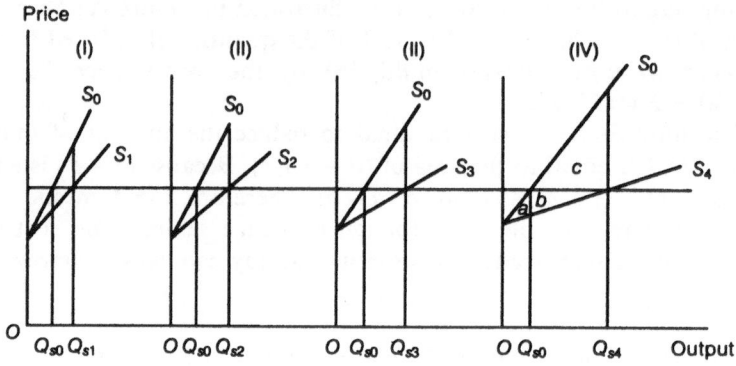

Figure A1.7 A trade-off function in input subsidies

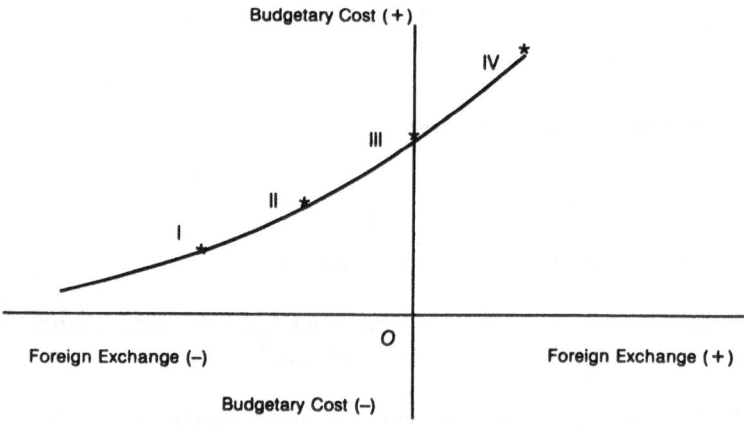

trade-off is described in Figure A1.7 in which a positive sign represents cost to the budget (BC) and earning for foreign exchange (FE) but a negative sign represents earning to the budget and outlay to foreign exchange.

When input subsidies are at the first two levels of (I) and (II) the country is in a food deficit position but when input subsidy reaches the level of (III) the country becomes totally self-sufficient. Further increase in input subsidy beyond level (III) makes the country become a net exporter as indicated at level (IV) (in Figure A1.7, at level IV, the country is earning foreign exchange through export).

A1.5.4 Implications of input subsidy

The analysis in A1.5.2 and A1.5.3 is based on a number of assumptions which are not very realistic. In addition, the issue of income and welfare redistribution is not addressed.

The norm of economy efficiency is based on free trade and perfect competition. Anything away from this norm is considered to involve welfare losses or the dead-weight losses. This precondition is prone to the following criticisms.

(a) Free trade and perfect competition recalls the arguments against Pareto Optimality which ignores the problem of trade-off between redistribution and efficiency.

(b) Assumption of perfect competition implies that the additional resources drawn from other activities due to input subsidy only represent the opportunity costs of those activities. One argument for government intervention is that the market may not work perfectly.

(c) The analysis also assumes full employment and no excess capacity of production. It implies that the resource costs $(a + b)$ are covered by market receipts, but are also paid by the subsidy. The resource cost (c) is paid by the budget but is not recovered from market receipt by the producers. If the government puts more weight on the welfare of producers than that of taxpayers and if full employment in the agricultural sector does not exist, than input subsidy may achieve the desired results expected by the government: fairer income distribution and more on-farm employment.

However, for an input subsidy to be successful in raising farm incomes, it needs to be associated with the pricing policy on products. In pre-reform China, as product prices were very low, input subsidy was used to compensate part of production taxation rather than as a means to increase farmers's incomes. In addition, subsidies need to be avoided for inputs for which the supply is relatively inelastic.

Input subsidies may prove particularly attractive if there are specific farming practices a government wishes to encourage, such as land reclamation or improved seeds; there may then be longer run social benefits to the community. In China, agricultural research and extensions, including the breeding of HYVs of rice, wheat and cotton, have been heavily subsidized by the state. State subsidies enabled the agricultural research and experimental stations throughout the country to help farmers to increase crop productivity over time. This is a very valuable experience for many similar developing countries to follow.

Input subsidies also have the advantage that they may provide a convenient way of concentrating income support on a particular group of farmers, such as hill farmers or remote communal farmers. It may be politically more feasible to aid disadvantaged farming groups by subsidizing investment on hill or small farms in the remote regions rather than by some other means such as PTP. Similarly, there are less likely to be adverse effects in factor and product markets when only a small proportion of farmers is affected.

On the other hand, a generalized input subsidy scheme is likely to help most the larger and prosperous farmers: the proportion of purchased inputs tends to increase with farm size.

A1.6   EXPORT TAXATION

A1.6.1   Introduction

Control over border trade by governments includes a complex set of policy instruments in terms of both price and quantity restrictions. Price control on border trade may consist of the following issues: foreign exchange regulations (such as deliberate overvaluation of domestic currency), import and export tariffs. Quantity control on border trade includes import/export quotas. To effectively control border trade, most governments exercise monopoly power in imports and exports through marketing boards or parastatals. In China, agricultural import and export activities are entirely controlled by the MOFERT (Chapter 6). With its monopoly power in trade, the government is able to control foreign exchange and impose taxes/ subsidies on exportable commodities. Before 1980, agricultural exports in China were subject to heavy taxes in the form of currency overvaluation. Since 1980, because the local currency has been gradually devalued, the degree of export taxation on agricultural goods must have been declining. Currency devaluation has been one effective method of stimulating agricultural exports which, in turn, have helped to diversify domestic agricultural production.

The objectives of export taxation can be summarized as follows:

(a) to increase government revenue;
(b) to stabilize domestic prices for exportable commodities;
(c) to protect the interest of domestic consumers;
(d) for countries (large countries) which have a dominant position in the trading of the commodities under concern, one additional objective may be to maximize domestic social welfare by using their monopolistic or oligopolistic power in the international markets.

The consequences of export taxation are obvious:

(a) domestic consumers will benefit because domestic consumption will be higher and prices lower for those exportables;
(b) domestic production will be discouraged and producers of such crops will suffer due to depressed production at lower prices;
(c) the government gains due to export taxes;
(d) total foreign exchange earning will be reduced because of depressed exports.

There are two ways of imposing export taxation.

(a) *Variable export levy.* This means that government or state parastatals pay the domestic producers a fixed price which on average is lower than the international (export parity) price. The amount of export tax per unit of product is various because the export parity price may be fluctuated due to the changes in the international markets. At a particular point of time, the export levy is equal to the difference of the fixed domestic price and the variable export parity price.
(b) *Fixed unit export levy.* This means that the export tax is constant regardless of the changes in the international markets. The domestic prices to the producers have to be changed if the export parity prices change.

Variable export levy has the advantage of ensuring a stable price for the domestic producers but suffers from the disadvantage of being unable to guarantee a stable tax revenue for the government. A fixed unit export levy can guarantee a stable revenue for the government but it cannot provide a stable price for the domestic market. It also requires that domestic prices be adjusted as frequently as prices in the international markets change.

The following examples assume that both international price (export parity) and export levy are constant. Two country cases are also studied in order to distinguish the different consequences of imposing export taxation on both a small and a large country. The welfare analysis is based on the assumption of free trade and perfect competition. Any intervention, in this case export taxation, is supposed to violate this efficiency norm and therefore incurs efficiency losses.

A1.6.2 Export taxation: a small country case

The assumption for a small country case is that the amount of production and export does not affect international prices. The country is a price-taker facing a horizontal demand curve which is perfectly elastic with respect to the international price. Thus, the demand curve for the export of the country is the world price P_w as shown on the left hand panel of Figure A1.8.

(1) Domestic production, consumption and export without export taxation

The left hand panel of Figure A1.8 illustrates the domestic demand and supply functions, represented by D and S respectively. Without export

Figure A1.8   Export taxation: a small country case

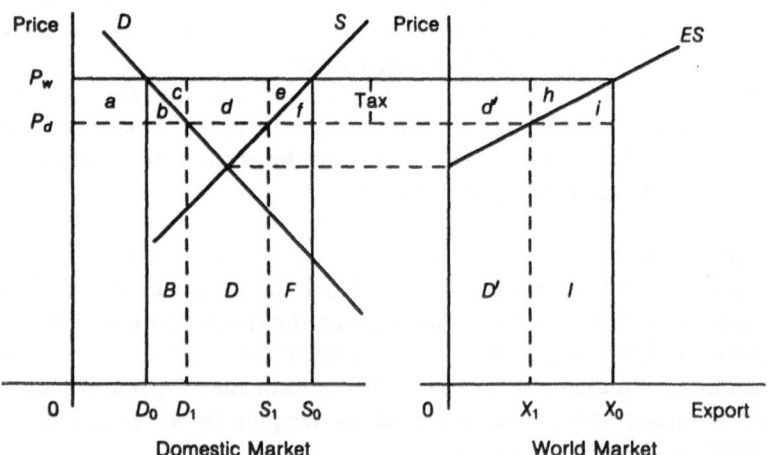

Domestic Market   World Market

taxation, both producers and consumers face the same international price, P_w, at which domestic production is S_0 and demand D_0. Total export is the difference between S_0 and D_0, i.e. $(S_0 - D_0)$. Total export can also be measured by the excess supply curve ES and P_w (international demand curve) on the right hand panel as X_0: $X_0 = (S_0 - D_0)$.

(2) Domestic production, consumption and export with export taxation

Suppose export taxation reduces the domestic price from P_w to P_d, domestic demand will increase from D_0 to D_1 and domestic supply will decrease from S_0 to S_1. The total amount of export will be curtailed from $(S_0 - D_0)$ or X_0 to $(S_1 - D_1)$ or X_1 as a result.

(3) Welfare effect of export taxation

Compared with the scenario of no export taxation, export taxation has the following welfare effects:

(a) Producer surplus is reduced by the amount of $(a + b + c + d + e)$.
(b) Consumer surplus is increased by the amount of $(a + b)$.
(c) The state budget gains by the amount of (d).
(d) Net social welfare loss is equal to $(c + e)$.

The loss of c is due to excessive demand and the loss of e is due to depressed supply after the imposition of export taxation.

(4) Domestic cost and foreign exchange earning with export taxation

Export taxation has the following consequences on export and domestic costs:

(a) Total foreign exchange earning is reduced by $(h + i + I)$.
(b) The reduction of domestic production reduces the domestic cost of production by $(f + F)$.
(c) Due to increased domestic consumption, consumer welfare increases by $(b + B)$.
(d) Total welfare loss due to reduced export
    $= (f + F) + (b + B) - (h + i + I)$.

Because $(D_0 - D_1) + (S_0 - S_1) = (X_0 - X_1)$, then $B + F = I$, $b + f = i$ and $c + e = h$, so total welfare loss due to reduced export is equal to:

$$(f + F) + (b + B) - (h + i + I) = -h = -(c + e),$$

which is exactly equal to the net social welfare loss in (3).

(5) A remark on social welfare losses

Both (3) and (4) reach the same conclusion on the net social welfare loss although the approaches used to calculate the loss are different. In (3), the approach is based on the domestic market and the welfare distribution among different interest groups. The loss occurs because the total gain to consumers and taxpayers (budget) is less than the total loss to producers. The net social loss is therefore explained by the sub-optimal consumption and production levels in the domestic market.

In (4), the approach is based on the trade-off between the change in domestic cost of production plus the change of domestic consumer gain and the change of foreign exchange earning. In this case, the loss is explained by the fact that the total amount of foreign exchange loss is not wholly recovered by the total sum of the cost saved in domestic production and the increase in domestic consumer surplus.

A1.6.3 Export taxation: a large country case

In this section, the effects on domestic production, consumption and welfare are based on the assumption that the country is large enough to influence the world price through its own export. In other words, the country is facing a downwards-sloping demand curve for its export. Countries which can influence the world price include Brazil (coffee), Thailand (rice) and Malaysia (rubber), Zimbabwe (tobacco), etc. The framework in Figure A1.9 is based on the Thai rice market (Wong, 1978).

(1) Effects of export taxation on production, consumption and export

In Figure A1.9, an export premium (tax) reduces the domestic price from P_w (export parity price without tax) to P_d but raises the international price up to P'_w. The international price is increased due to the export taxation of the country concerned because the international demand curve for the product is not perfectly elastic. The effects of export taxation on domestic production, consumption and export can be summarized as follows.

Figure A1.9   Export taxation: a large country case

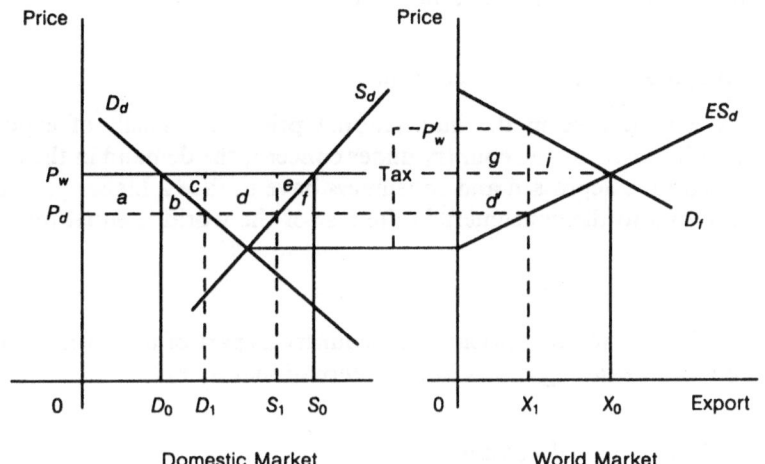

Domestic Market   World Market

(a) domestic supply is reduced from S_0 to S_1;
(b) domestic consumption increases from D_0 to D_1;
(c) total export is reduced from $(S_0 - D_0)$ to $(S_1 - D_1)$ or from X_0 to X_1. $(S_0 - D_0) = X_0$, and $(S_1 - D_1) = X_1$.

(2) Welfare effect of export taxation on the country under study

As in the small country case, domestic producers are still the net losers while domestic consumers and taxpayers are the beneficiaries of export taxation. However, the total benefit to domestic consumers and taxpayers may not be necessarily less than the total loss to domestic producers. In other words, the country under study may not incur any net social welfare losses due to export taxation. On the contrary, the country under study may become a net winner by imposing such a tax. This is because the international price can be raised above the initial equilibrium level due to reduced supply to the world market. The welfare effect on the country can be calculated as follows.

(a) Loss to domestic producer surplus $(a + b + c + d + e)$.
(b) Gain to domestic consumer surplus $(a + b)$.
(c) Gain to government budget (export tax) $(d + g = d' + g)$.
(d) Net total welfare loss (gain)
$$(a + b + d + g) - (a + b + c + d + e) = g - (c + e).$$

If $g < (c + e)$, the country incurs a net loss of social welfare; but if $g > (c + e)$, the country has a net gain.

(3) Welfare effect on the rest of the world

Due to an increase in the international price as a result of export taxation imposed by the country under concern, the demand in the rest of the world is depressed and consumers have to face a higher price at P'_w. The loss to the consumers in the rest of the world is equal to:

$$(g + i)$$

where g is captured by the exporting country as part of tax revenue but i becomes a deadweight loss in the international market.

(4) Effect on global welfare

The net social welfare loss to the whole world including the country under concern and the rest of the world is equal to:

$$(g + i) - [g - (c + e)] = (i + c + e)$$

A1.6.4 Implications of export taxation

Export taxes are one major source of government revenue in many developing countries, especially in those countries which have to rely heavily on agricultural export to finance domestic investment and government expenditure.

In a small country, because the amount of export by the country cannot affect the world price, there will be an inevitable net loss of social welfare. In a large country, because the amount of its export can influence the international price, the country concerned may be a net gainer rather than a loser in terms of domestic social welfare in the short run. However, even in a large country, there are still a number of long run undesirable effects of export taxation (these effects are also applicable to the small countries).

(a)   Importers of the commodities may turn to other exporters or increase their self-sufficiency. This will depress the demand for the exports by the countries imposing export levies.

(b) The export premium represents a significant burden on domestic producers. This will inevitably lead to depressing producers' incentives to improve yields by adopting new technologies and applying more modern inputs such as fertilizers and improved seeds. Thus, producers may fail to shift to a higher production frontier. It may be because of this that many poor countries have failed to improve farm productivity, especially for the small holders.

(c) Reduced export due to taxation will lead to losses of foreign exchange earning and damage the balance of payments.

(d) Loss of employment opportunities for products under controls. If export crops are more labour intensive than other crops, resource re-allocation from exportables to non-exportables will reduce the overall capacity to absorb surplus labour in the rural areas.

A1.7   SUMMARY

A number of simple frameworks are presented in this chapter, partly because the policy instruments have been implemented in China and many other developing countries. All the mechanisms demonstrate the general direction of impacts on production, consumption, and the welfare of consumers and producers. Some policy instruments also involve significant budgetary costs (in cases of subsidy) or revenues (in cases of taxation).

Apart from the compulsory production quota, all the mechanisms are discussed at the 'national' level. The impacts of different policy instruments at the regional level, however, may be far more complicated, although final conclusions may be still similar. In addition, graphical analysis cannot quantify the impacts of different policy instruments.

To quantify the effects and understand the impacts of policy instruments at the regional level, a mathematical model has to be employed. Therefore, the rest of the book will be devoted to constructing some workable mathematical models for the Chinese grain sector.

A2 Policy Models for the Chinese Grain Sector

A2.1 INTRODUCTION

In Chapter A1, policy analyses are graphically presented as a general case. Two important theorems are also proven in the respective two-dimensional diagrams. The exact quantitative effects of any policy, however, have to rely on empirical studies. This chapter develops a number of policy models special for the Chinese grain sector. On the one hand, the models are constructed in such a way that empirical data can be used to test the theorem presented in the previous chapter. On the other, they will enable us to produce some useful policy simulation results concerning the most important agricultural policy instruments in China.

Cross-sectional data collected for 1985 represent the turning point of policy during economic reforms (see Chapter 4). Data manipulation and the estimations of regional demand functions, supply prices or production costs, transportation costs, and definition of geographical regions will be discussed in Chapter A3. Computational results are presented in Chapter A4. Conclusions and policy implications are discussed in Chapter 7. This chapter will only present the mathematical formulation and solution interpretations for the policy models.

The framework of all the models in this chapter is that of the quadratic spatial equilibrium. The principle of a spatial equilibrium model is to adopt the Marshall (1920), Hicks (1939) and Samuelson (1948) neoclassical system, where under appropriate assumptions aggregate community demand and supply functions may be derived from the underlying community utility and total production or cost functions. This framework, along with the concept of a quasi-welfare function, or social pay-off function, generated along the lines of producers' and consumers' surplus, will form the basis for the specification of spatial models where all quantities and prices are the unknowns to be determined.

In a spatial equilibrium model, the objective is to maximize a net social pay-off which is equal to the total social pay-off minus total transportation costs. The net social pay-off from trade for an exporting region is equivalent to the conventional definition of producer surplus accruing to producers. For importing regions the net social pay-off is akin to the additional consumer surplus.

Thus in relation to Figure A2.1, in the absence of transshipment costs the net social pay-off from trade to both partners (as opposed to no trade) is denoted by subtracting the area OP'_1AQ from OP'_2AQ leaving a net social pay-off from trade of $P'_1AP'_2$ of which P'_1AP accrues to region 1 (the exporter) and P'_2AP to region 2 (the importer). However if there are transshipment costs of t_{12} total inter-regional transshipment will decrease from x'_{12} to x^*_{12} and the net social pay-off is the sum of the two triangular areas $P'_2BP^*_2$ and $P^*_1CP'_1$.

Figure A2.1   Excess demand and social pay-off

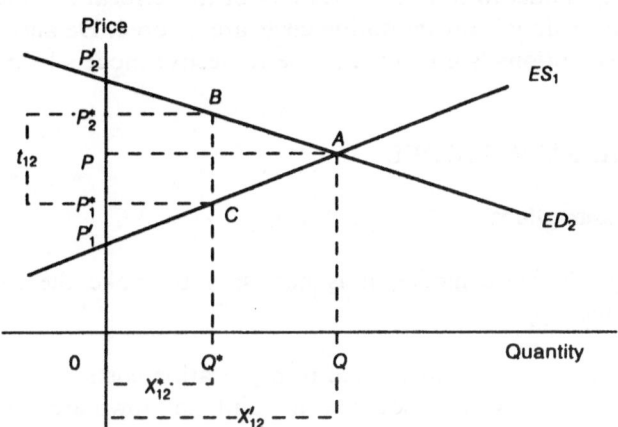

If the regional demand and supply functions are known and denoted as $f(d_i)$ and $f(s_i)$ for region i $(i = 1, 2, \ldots, n)$ and if the inter-regional transportation costs are fixed at t_{ij} $(i, j = 1, 2, \ldots, n)$, the net social pay-off for a multi-regional case can be represented by the following equation:

$$NW(s, d, x_{ij}) \equiv \sum_i^n \int_{d_i}^{d'_i} f(d_i)d(d_i) - \sum_i^n \int_{s_i}^{s'_i} f(s_i)d(s_i)$$

$$- \sum_i^n \sum_j^n t_{ij}x_{ij} \tag{A2.1}$$

where $d_i{}'$, s_i' are respectively the pre-trade demand and supply in region i, and x_{ij} is the transshipment from region i to region j. d_i^* and s_i^* are the after-trade equilibrium demand and supply in region i. The principle in a spatial equilibrium model is to maximize this objective function, subject to constraints, in order to obtain the optimal demands, supplies, inter-regional flows and prices for all individual regions. If both $f(d_i)$ and $f(s_i)$ are functions of prices, we have two sets of unknown variables, prices and quantities which enter multiplicatively so that the objective function is quadratic.[1]

In the light of data availability and the objectives of this study of the Chinese grain sector, the above model is modified in a number of ways. Firstly, the supply functions are assumed to be horizontal for all the individual regions. Secondly, an upper bound constraint is imposed on every policy model to reflect the scarcity of this critical resource. And lastly, intra-regional transportation costs are ignored. Detailed discussion of assumptions is presented in the respective models below.

A2.2 THE BASIC MODEL

A2.2.1 Assumptions

To set up the basic model, it is necessary to make the following assumptions.

(a) There are no intra-regional transportation costs.

(b) There is only one resource constraint, i.e., sown areas available for grain production. In this case the land constraints are artificially set to be 1.04 times those of the actual areas sown to grain crops in 1985 for all regions. The reason why we introduce land constraints and why we use 1.04 times the actual crop areas is explained later.

(c) There are 26 regions which are separated by a transportation cost per physical unit which is independent of volume. There are no legal restrictions to limit the actions of the profit-seeking traders in each region.

(d) All regions have the same price 'coefficient' in demand for per capita consumption of grain; the intercepts of the demand functions are estimated with a linear equation with regard to actual consumption in 1985. Detailed explanations and estimations of demand functions for all regions are given in Chapter A3.

(e) Market supply prices, which may be considered as production costs, are assumed to be exogenous because the supply functions cannot be estimated with available data. Supply prices or production costs are also estimated in Chapter A3.

(f) Grain in the model refers to the aggregate of all kinds of grains integrated together, including rice, wheat, corn, millet, sorghum and tuber crops (such as sweet potatoes). In China, tuber crops have been considered as grain and they are counted at a ratio of 5 to 1, i.e., 5kg of sweet potatoes is equivalent to 1kg of rice or wheat. It is assumed that grain is homogeneous in terms of demand and supply for all regions. The grain demand for animals, seeds, local storage and other miscellaneous purposes is estimated for each region and treated as a necessary (or fixed) demand.

(g) Imports and exports on the national borders are ignored, i.e., we treat the grain market of China as a closed market.

A2.2.2 Model definitions[2]

(a) The demand functions

Let P_{di} and d_i represent the demand price and quantity demanded in region i, then the demand function in region i can be expressed as:

$$P_{di} = h_{0i} - h_{1i}d_i \quad \text{for all } i \tag{A2.2}$$

(b) Production costs

The production cost in region i is denoted as P_{si} (for all i), which is treated as a constant in line with the assumptions above.

(c) Unit transportation costs and inter-regional flows

The transportation cost between regions i and j is denoted as t_{ij}, for all i and j ($t_{ij} > 0$).

The quantity of inter-regional trade between regions i and j is denoted as x_{ij}, for all i and j ($x_{ij} \geq 0$).

(d) Quantity-flow constraints

For each region, total supply plus inflows from other regions must be greater or equal to the total demand plus the outflows of that region to all the other regions, i.e.,

$$s_i + \sum_{j \neq i} X_{ji} - d_i - \sum_{i \neq j} X_{ij} \geq d_i^0 \quad \text{for all } i \qquad (A2.3)$$

where d_i^0 is the fixed demand (necessary demand) in region i. It may include the demand accounted for by animals, seed requirement, losses and local necessary storage or other uses. s_i is the total supply in region i. $\sum_{j \neq i} X_{ji}$ is the sum of total inflows from all regions except region i itself to region i. d_i is the total demand in region i. $\sum_{i \neq j} X_{ij}$ is the sum of outflows to all regions except region i itself from region i.

(e) Land constraints

Given an average unit land output of grain in each region and the total land available for growing grain crops, the following condition is imposed:

$$r_i s_i \leq L_i \quad \text{for all } i \qquad (A2.4)$$

where, r_i is a known constant: it is the inverse of the average unit land output of grain in region i. And L_i is the maximum land available for grain production in region i.

There are two reasons why this constraint is necessary. Firstly, because a complete model, which includes all the competitive farm crops together, cannot be built, the non-grain crops are allocated a certain proportion of total arable land. Secondly, since we cannot construct regional supply functions due to non-availability of data, we assume a constant production cost (supply price) for each region. Therefore, the regional supply curve is perfectly elastic with respect to a constant price as shown in Figure A2.2(a), although the national supply curve may still be upward-sloping as shown in Figure A2.2(b).

If we do not impose the upper bound constraints on land, the low-cost regions may produce indefinitely while the high-cost regions may not produce anything. This would be unrealistic. Thus, the introduction of upper bound land constraints is partly to combat the shortcomings of the assumption of a constant production cost.

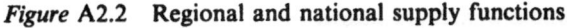

Figure A2.2   Regional and national supply functions

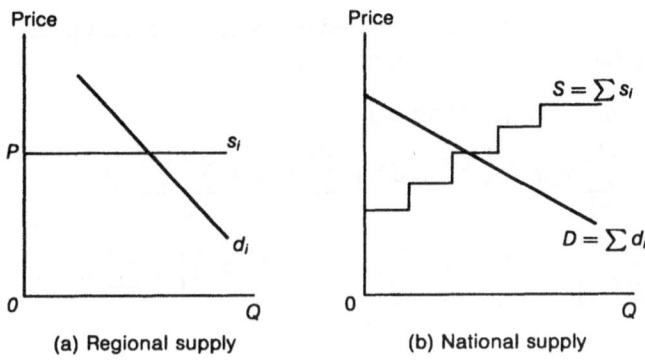

(a) Regional supply   (b) National supply

The land availability for grain is based on the actual sown areas in 1985 for all the regions. However, to allow for some flexibility, the upper bound land constraints are set to be 1.04 times those of the actual areas in 1985. The choice of a 4 per cent increase over the actual areas is arbitrary.

A2.2.3 Mathematical formulation

Given the definitions made above, the objective function of the basic model is given as follows:

$$\text{Max.} \sum_i NW(d_i, s_i, x_{ij}) = \sum_i h_{0i} d_i - \tfrac{1}{2} \sum_i h_{1i} d_i^2 - \sum_i p_{si} s_i$$
$$- \sum_{i \neq j} \sum_{j \neq i} t_{ij} x_{ij} \qquad (i, j = 1, 2, \dots, 26)$$
$$(A2.5)$$

And with this objective function, the following problem is set:

Problem A2.1: find $(d_i^*, s_i^*$ and x_{ij}^* for all i and $j)$ that maximizes (A2.5) subject to:

$$s_i + \sum_{j \neq i} x_{ji} - d_i - \sum_{i \neq j} x_{ij} \geq d_i^0 \qquad \text{for all } i$$
$$r_i s_i \leq L_i \qquad \text{for all } i$$
$$s_i, d_i, x_{ij} \geq 0 \qquad \text{for all } i \text{ and } j \qquad (A2.6)$$

A2.2.4   Economic interpretations

The following Lagrangean is formed by using (A2.5) and (A2.6) and by changing the problem to a minimization one.

$Z(d_i, s_i, x_{ij}, e_i, q_i, \text{ for all } i, j)$

$$= -\sum_{i=1}^{n} h_{0i}d_i + \frac{1}{2}\sum_{i=1}^{n} h_{1i}d_i^2 + \sum_{i=1}^{n} P_{si}s_i + \sum_{i \neq j}\sum_{j \neq i} t_{ij}x_{ij}$$

$$+ \sum_{i=1}^{n} e_i\left[d^0 - \left(-d_i - \sum_{i \neq j} x_{ij} + s_i + \sum_{j \neq i} x_{ji}\right)\right] + \sum_{i=1}^{n} q_i(r_i s_i - L_i)$$

$$\tag{A2.7}$$

where e_i and q_i (for all i) are the respective Lagrangean multipliers for the constraints.

The optimality conditions (first-order Kuhn-Tucker conditions) are as follows:

(a) $\partial Z^*/\partial d_i = -h_{0i} + h_{1i}d_i^* + e_i^* \geq 0$
     and $(\partial Z^*/\partial d_i)d_i^* = 0$;

(b) $\partial Z^*/\partial s_i = P_{si} - e_i^* + r_i q_i^* \geq 0$
     and $(\partial Z^*/\partial s_i)s_i^* = 0$;

(c) $\partial Z^*/\partial x_{ij} = t_{ij} - (e_j^* - e_i^*) \geq 0$
     and $(\partial Z^*/\partial x_{ij})x_{ij}^* = 0$;

(d) $\partial Z^*/\partial e_i = d_i^0 - (-d_i - \sum_{i \neq j} x_{ij} + s_i + \sum_{j \neq i} x_{ji}) \leq 0$
     and $(\partial Z^*/\partial e_i)e_i^* = 0$;

(e) $\partial Z^*/\partial q_i = r_i s_i^* - L_i \leq 0$
     and $(\partial Z^*/\partial q_i)q_i^* = 0$;   (A2.8)

Economic interpretations are given as follows.

(1) Optimum consumption

By interpreting the Lagrangean e_i^* as the optimum market demand price in region *i* for the product and combining this with the second part of (A2.8a), we can give (A2.8a) the following economic interpretation: at the optimum: (i) when the consumption in region *i* d_i^* is positive, the regional demand price is equal to the market demand price e_i^*; and (ii) when $d_i^* = 0$, the market demand price must be greater than or equal to the regional demand price.

(2) Optimum production (supply)

The first part of (A2.8) can be written as

$$P_{si} + r_i q_i^* \geq e_i^* \quad \text{for all } i \tag{A2.9}$$

and e_i^* can be interpreted as the optimum market supply price in region i. P_{si} is the production cost in the same region.

The first part of (A2.8e) can be written as

$$r_i s_i^* \geq L_i \quad \text{for all } i \tag{A2.10}$$

If q_i^* is interpreted as the reduced cost or shadow price of land in region i, combining (A2.9) and (A2.10) together, then (A2.8b) and (A2.8e) can be interpreted as follows: (i) when supply $s_i^* = 0$, the market supply price must be less than or equal to the regional supply price (cost); (ii) if $q_i^* > 0$, then $L_i = r_i s_i$, or the available arable land for grain in region i is fully utilized. And from (A2.9), $P_{si} + q_i^* r_i = e_i^*$. Hence the regional production cost is smaller than the market supply price in region i. (iii) If $s_i > 0$ and $q_i^* = 0$, then the market supply price must be equal to the regional production cost. The condition for this is $L_i \geq r_i s_i^* > 0$. Thus,

(1) if $s_i^* = 0$ $(q_i^* = 0, L_i > r_i s_i^*)$, $P_{si} > e_i^*$ $\quad (\geq 0)$
(2) if $s_i^* > 0$ $(q_i^* > 0, L_i = r_i s_i^*)$, $P_{si} + q_i^* r_i = e_i^*$ (≥ 0)
(3) if $s_i^* > 0$ $(q_i^* = 0, L_i \geq r_i s_i^*)$, $P_{si} = e_i^*$ $\quad (\geq 0)$ $\quad$ (A2.11)

(3) Optimum flows, optimal excess demand and supply

In conditions (A2.8c) and (A2.8d), when e_i^* is positive, the total supply s_i^* plus total inflows from other regions $\sum_{j \neq i} x_{ji}$, minus the sum of total demand d_i^* plus total outflows to other regions $\sum_{i \neq j} x_{ij}$, must be equal to the local fixed demand d_i^0. If $e_i^* = 0$, the difference between the former and the latter will be greater than or at least equal to d_i^0. Thus,

(1) if $e_i^* > 0$, $-d_i^* + s_i^* - \sum_{i \neq j} x_{ij}^* + \sum_{j \neq i} x_{ji}^* = d_i^0$
(2) if $e_i^* = 0$, $-d_i^* + s_i^* - \sum_{i \neq j} x_{ij}^* + \sum_{j \neq i} x_{ji}^* \geq d_i^0$ $\quad$ (A2.8d)

A2.2.5 Two comments about the model

Since we assume that all regional supply prices (costs) entering the model are constant, all the supply curves are horizontal, or perfectly elastic with respect to the costs (see Figure A2.2a). A regional supply price may also be regarded as the minimum long-term average cost of production in a particular region. And such a cost must be greater than zero. Therefore, according to condition (A2.8b), as long as $P_{si} > 0$, the market equilibrium price e_i^* (for all i) is always greater than zero.

With this special feature, we can draw two general comments about this model.

(i) $-d_i^* + S_i^* - \sum_{i \neq j} x_{ij}^* + \sum_{j \neq i} x_{ji}^* = d_i^0$ must be held for all i in the final solution. This states that the possibility of excess demand or excess supply in any region is eliminated.

(ii) There is a possibility that some deficit regions might not produce anything if the equilibrium market prices (e_i^*) happened to be smaller than the regional production costs (P_{si}). A zero production solution to a high-cost region may be unrealistic. If we could construct the usually upward-sloping supply curves, the possibility of zero-production regions would be unlikely to exist in the models in which each region consists of a large province. Therefore, a horizontal supply curve is quite naive and unrealistic. Nevertheless, because we cannot obtain estimates of the more realistic upward sloping supply curves, we have to introduce such a strong assumption. Apart from this, as we have mentioned before, our models are very partial: they do not include other crops, time dimension, imports and exports, etc. Thus, policy analysis and conclusions, based on a partial structure of the economy, must inevitably rely more heavily on 'ceteris paribus' conditions as implied in the above assumptions. This is a dilemma inevitable in economic science, especially when we are confronted with severe lack of data.

A2.3 BASIC MODEL WITH ACREAGE QUOTA

A2.3.1 Definitions

In order to analyse the welfare effects on producers, consumers, and taxpayers, and the effects on the patterns of grain production of a compulsory production quota, this section establishes the so-called 'quota' model.

The assumptions in this section are exactly the same as those listed in Section A2.2 except that a lower-bound land constraint is introduced as a means of imposing a production quota for each region.

The actual model sets the limits to be 90 per cent of the total actual areas sown to grain crops in 1985 for all the regions. I_i is used to denote this limit in region i. As land productivity is treated as a constant, a lower bound on land area is equivalent to a lower bound on production.

A2.3.2 Mathematical formulation

The objective function of this model is exactly the same as (A2.5). Therefore, the following problem can be formed as:

Problem A2.2: find (d_i^*, s_i^*, and x_{ij}^*, for all i and j) that maximizes (A2.5) subject to:

$$s_i - d_i + \sum_{j \neq i} x_{ji} - \sum_{i \neq j} x_{ij} \geq d_i^0 \quad \text{for all } i$$
$$r_i s_i \leq L_i \quad \text{for all } i$$
$$r_i s_i \geq I_i \quad \text{for all } i$$
$$s_i, d_i, x_{ij} \geq 0 \quad \text{for all } i \text{ and } j \quad \text{(A2.12)}$$

A2.3.3 Economic interpretations

(1) The optimality conditions

The following Lagrangean is formed by using (A2.5) and (A2.12) and by changing the problem to a minimization one.

$$Z(d_i, s_i, x_{ij}, e_i, q_i, u_i, \text{ for all } i, j) = -\sum_{i=1}^{n} h_{0i} d_i + \frac{1}{2} \sum_{i=1}^{n} h_{1i} d_i^2$$
$$+ \sum_{i=1}^{n} P_{si} s_i + \sum_{i \neq j} \sum_{j \neq i} t_{ij} x_{ij} + \sum_{i=1}^{n} e_i \left[d^0 - \left(-d_i - \sum_{i \neq j} x_{ij} + s_i + \sum_{j \neq i} x_{ji} \right) \right]$$
$$+ \sum_{i=1}^{n} q_i (r_i s_i - L_i) + \sum_{i=1}^{n} u_i (I_i - r_i s_i) \quad \text{(A2.13)}$$

where e_i, q_i, and u_i (for all i) are the respective Lagrangean multipliers for the constraints.

The optimality conditions (first-order Kuhn–Tucker conditions) are as follows:

(a) $\partial Z^* / \partial d_i = -h_{0i} + h_{1i} d_i^* + e_i^* \geq 0$
    and $(\partial Z^* / \partial d_i) d_i^* = 0$;

(b) $\partial Z^* / \partial s_i = P_{si} - e_i^* + r_i q_i^* - r_i u_i^* \geq 0$
    and $(\partial Z^* / \partial s_i) s_i^* = 0$;

(c) $\partial Z^* / \partial x_{ij} = t_{ij} - (e_j^* - e_i^*) \geq 0$
    and $(\partial Z^* / \partial x_{ij}) x_{ij}^* = 0$;

(d) $\partial Z^* / \partial e_i = d_i^0 - (-d_i - \sum_{i \neq j} x_{ij} + s_i + \sum_{j \neq i} x_{ji}) \leq 0$
    and $(\partial Z^* / \partial e_i) e_i^* = 0$;

(e) $\partial Z^* / \partial q_i = r_i s_i^* - L_i \leq 0$
    and $(\partial Z^* / \partial q_i) q_i^* = 0$;

(f) $\partial Z^* / \partial u_i = I_i - r_i s_i^* \leq 0$
    and $(\partial Z^* / \partial u_i) u_i^* = 0$.   (A2.14)

(2) Economic interpretations

All the explanations of (a), (c) and (d) of (A2.14) are the same as those in problem A2.1. Economic interpretations here will only be given to (b), (e) and (f).

The first part of (b), (e) and (f) of (A2.14) can be written as:

$$P_{si} + r_i q_i^* - r_i u_i \geq e_i^* \qquad \text{for all } i \qquad\qquad (A2.15)$$

$$L_i - r_i s_i^* \geq 0 \qquad \text{for all } i \qquad\qquad (A2.16)$$

$$r_i s_i^* - I_i \geq 0 \qquad \text{for all } i \qquad\qquad (A2.17)$$

e_i^* is also interpreted as before as the optimum market price in region i. q_i^* and u_i^* can be interpreted as the optimum reduced costs (shadow prices) of land in region i with respect to the upper and lower bound constraints of land respectively.

Combining with the last parts of (A2.14b), (A2.14c) and (A2.14f), the optimum production in each region can be interpreted as follows:

(1) When $q_i^* > 0$, then $L_i = r_i s_i^* > I_i$, so $u_i^* = 0$. The available arable land for grain in region i is fully utilized. And by (A2.14b), $P_{si} + r_i q_i^* = e_i^*$. This implies that the optimum market

price is greater than the regional production cost, with the difference measured by $r_i q_i^*$.

(2) When $u_i^* > 0$, then $I_i = r_i s_i^* < L_i$, so $q_i^* = 0$. And by (A2.14b), $P_{si} - r_i u_i^* = e_i^*$. This implies that the regional production cost is greater than the market equilibrium price e_i^*, with the difference measured by $r_i u_i^*$ i.e. the opportunity cost of imposing an extra unit of output in the region.

(3) When $q_i^* = u_i^* = 0$, then $I_i \leq r_i s_i \leq L_i$. And by (A2.14b), $P_{si} = e_i^*$. This implies that the regional production cost is exactly equal to the market equilibrium price in that region.

(4) As a last point about regional supply, it is obvious from (1) and (2), that q_i^* and u_i^* cannot be both positive in the same region as long as $L_i \neq I_i$. Thus

$$\text{if } q_i^* > 0 \ (u_i^* = 0, \ L_i = r_i s_i^*); \quad P_{si} + r_i q_i^* = e_i^* \ (> 0);$$
$$\text{if } u_i^* > 0 \ (q_i^* = 0, \ I_i = r_i s_i^*); \quad P_{si} - r_i u_i^* = e_i^* \ (> 0);$$
$$\text{if } q_i^* = u_i^* = 0 \ (I_i \leq r_i s_i^* \leq L_i); \ P_{si} = e_i^* \qquad (> 0). \quad \text{(A2.14)}$$

A2.4 MODELS FOR CONSUMER SUBSIDIES

In this section, two models concerning consumer subsidies, based on the previous models, will be established. All the assumptions made in section 2 are still valid in this section. Problem A2.3 will incorporate the consumer subsidies into the model set up in section 2 or problem A2.1. Problem A2.4 will incorporate the lower bound constraints on land into the model set up by problem A2.3.

A2.4.1 Problem definitions

(1) A simple analysis on consumer subsidies

The Chinese government has subsidized urban consumers for many items including grain consumption. Urban consumers usually purchase foodgrains from the state commercial departments at a very low price. As shown in Figure A2.3, suppose consumers buy grains from the state (distributed at local grain shops owned by the state), at P_c, which is supposed to be lower than the equilibrium market price.

Figure A2.3 shows the aggregate national demand and supply schedules. At equilibrium, with perfect competition, the optimal price and quantity (aggregated or national average) are P^* and Q^* respectively. With consumer subsidies, the consumer price is Pc and the total demand increases from Q^* to Q_c. Suppose it is a closed economy, i.e., there is no import to cover the difference between demand and supply, and suppose there are no restrictions on consumers' demand at the price level P_c, then producers have to produce at least Q_c, the producers' price (supply price) must increase from P_c to P_s. The following model, therefore, assumes that the state is willing to pay the difference between the consumer (demand) price and the producer (supply) price, i.e., $P_s - P_c$. As a result, the total state budgetary costs are equal to the area $P_sbcP_c = (P_s - P_c) \times OQ_c$.

Figure A2.3   Consumer subsidies

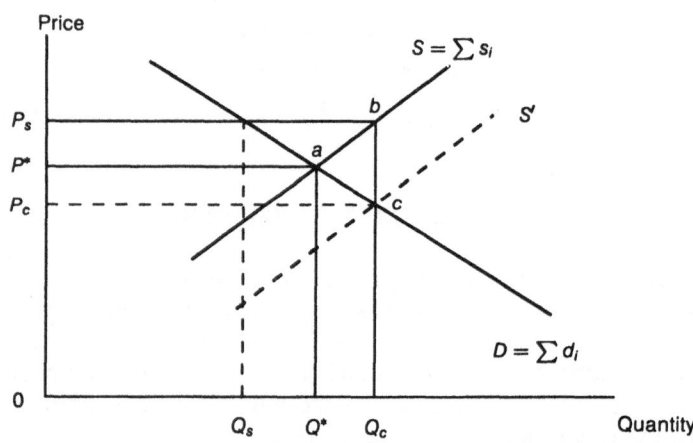

The welfare impact on consumers, producers and the society as a result of consumer subsidies is obvious. Both producers and consumers will gain at the expense of the state budget. Since the budgetary costs are greater than the sum of gains to producers and consumers, there is a net social welfare loss, which, as shown in the figure, is equal to the triangular area *abc*.

(2) Mathematical formulation

Suppose producers receive a subsidy which is equivalent to the difference between P_s and P_c, i.e. $(P_s - P_c)$, then the supply curve will shift from S to S' as shown in Figure A2.3. Consequently both demand and supply will be equal to Q_c with the equilibrium price P_c. But how can such changes be incorporated into the basic model described in problem A2.1? The answer depends upon whether P_c is known as a result of consumer subsidies. Since state subsidies are available only to the urban population, but not to the rural residents, there will be two consumer prices: one for the urban population, and the other for the rural population. The demand price of the urban people is determined by state subsidies, while the price for the rural people is determined by free market forces. Because the model described in section A2.1 is an aggregated model, i.e., it incorporates both the urban and the rural consumers for both the demand and supply, the new national equilibrium demand price with state subsidies to the urban consumers must be in between the free market price (for the rural people) and the subsidized price (for the urban people). The method and estimation of such a price is presented in the next chapter. Here we only consider the formulation of the model. Suppose the new price is P_c which is proportional to the former equilibrium price P_i^*, thus $P_c = \alpha P^*$ $(0 < \alpha < 1)$. Assuming that the total supply can be increased from Q^* to Q_c by reducing the production cost P_{si} $(i = 1, 2, \ldots n)$ with the same proportion $(1 - \alpha)P_{si}$ to P_{ci} $(i = 1, 2, \ldots n)$ for all regions, then the mathematical model can be formed by problem A2.3 as follows:

Problem A2.3: find $(d_i^*, s_i^*$ and x_{ij}^*, for all i and $j)$ that maximizes

$$\sum_i NW(d_i, s_i, x_{ij}) = \sum_i h_{0i}d_i - \tfrac{1}{2}\sum_i h_{1i}d_i^2 - \sum_i(\alpha P_{si})s_i - \sum_{i \neq j}\sum_{j \neq i} t_{ij}x_{ij}$$

or

$$\sum_i NW(d_i, s_i, x_{ij}) = \sum_i h_{0i}d_i - \tfrac{1}{2}\sum_i h_{1i}d_i^2 - \sum_i P_{ci}s_i - \sum_{i \neq j}\sum_{j \neq i} t_{ij}x_{ij} \qquad (A2.18)$$

subject to (A2.6)

where $P_{ci} = \alpha P_{si}$

(3) Subsidy model with lower-bound land constraints

If land lower bound constraints are introduced to problem A2.3, we can form the following problem:

Problem A2.4: find $(d_i^*, s_i^*$ and x_{ij}^*, for all i and $j)$ that maximizes equation (A2.18) subject to (A2.12).

A2.4.2 Economic interpretations

The optimality conditions and solutions of problems A2.3 and A2.4 are similar to those of problems A2.1 and A2.2 respectively. However, the impacts of consumer subsidies arranged in such a way as described by problems A2.3 and A2.4 will be different from those of problems A2.1 and A2.2. Since all the regions have upper bound constraints of land and the total outputs with consumer subsidies in problems A2.3 and A2.4 are expected to be greater than without subsidies in problems A2.1 and A2.2, the extra outputs will be more likely to be produced by those regions with relatively high production costs. The results of our computations and further analysis of the impacts on patterns of grain production will be presented in Chapter A4.

A2.5 MODELS FOR PRODUCTION TAXATION

In this section two models will be established on the assumption that producers have to deliver a certain amount of their outputs to the state at a low price (lower than the free market price). Again, the assumptions made in Section A2.2 remain valid in this section. The difference between the procurement price set for the amount delivered to the state and the free market price may be considered as the *producers' tax* forced upon the grain producers.

A2.5.1 Problem definitions

(1) A simple analysis of producers' tax

Usually there are a number of reasons why the government imposes some compulsory procurement from the farmers at a relatively low price: (a) to secure a certain amount of grain for the urban (or industrial) sector; (b) to extract resources out of agriculture for urban and industrial development; and (c) to maintain a certain degree of food self-sufficiency.

As discussed in Chapter 4, grain subsidies incurred substantial budgetary costs for the government. To reduce the cost, the government has imposed the so-called indirect taxes on producers.

The following models show how this policy affects prices, demands and supplies. As shown in Figure A2.4, without government intervention, the optimal equilibrium demand and supply will be the same as Q^* and the equilibrium price will be P^*. Suppose it is a closed economy, i.e., without exports and imports: the total supply will be reduced from Q^* to Q_r, while the market demand price will increase from P^* to P_d. If the tax imposed upon the producers is equivalent to the difference between P_d and P_r, the budgetary gains to the state will be equal to the rectangular area $P_dbcP_r = (P_d - P_r)(OQ_r)$. Both producers and consumers will lose, with producers losing an amount equal to the area P^*acP_r and consumers losing an amount equal to the area P_dbaP^*. Since the sum of losses to both the producers and the consumers is greater than the gains to the government, there will be a net loss of social welfare, which, in this case, is equal to the triangular area *abc*.

Figure A2.4   Production taxation

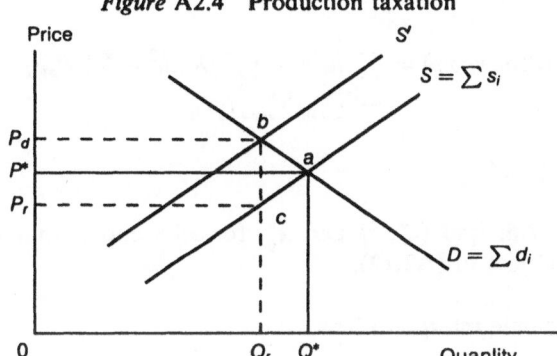

(2) Mathematical formulation of the models

The effect on supply or production by imposing a producers' tax (a low procurement price in this case) is the same as increasing the production cost in the basic model specified in problem A2.1. If the average production cost increases from P^* to P_d as a result of a tax, the supply curve will shift to the left from S to S' as shown in Figure A2.4. The new equilibrium price and supply are P_d and Q_r respectively. Suppose P_d is known to be the average national production cost and is defined as

$$P_d = \beta P^* \quad (\beta > 1) \tag{A2.19}$$

where β is a constant.

On the regional level, we assume that the total supply Q_r is reached if the original production costs for all regions P_{si} (for all i) increase by the same proportion: i.e.

$$P_{di} = \beta P_{si} \qquad\qquad (A2.20)$$

where P_{di} is the new production cost after taxation in region i which is so set that the total national production will be exactly equal to Q_r.

With these specifications and the assumptions, we formulate the following problems:

Problem A2.5: find (d_i^*, s_i^* and x_{ij}^*, for all i and j) that maximizes

$$\sum_i NW(d_i, s_i, x_{ij}) = \sum_i h_{0i}d_i - \tfrac{1}{2}\sum_i h_{1i}d_i^2 - \sum_i \beta P_{si}s_i$$
$$- \sum_{i\neq j}\sum_{j\neq i} t_{ij}x_{ij}$$

or

$$\sum_i NW(d_i, s_i, x_{ij}) = \sum_i h_{0i}d_i - \tfrac{1}{2}\sum_i h_{1i}d_i^2 - \sum_i P_{di}s_i$$
$$- \sum_{i\neq j}\sum_{j\neq i} t_{ij}x_{ij} \qquad (A2.21)$$

subject to (A2.6).

Problem A2.6: find (d_i^*, s_i^* and x_{ij}^*, for all i and j) that maximizes (A2.21), subject to (A2.12).

A2.5.2 Economic interpretations

The optimality conditions and solutions of problems A2.5 and A2.6 are similar to those of problems A2.1 and A2.2 respectively. The computation results and cost estimation for the vector P_{di} (for all i) will be given in the following chapters.

As shown in Figure A2.4, it is obvious that total production shrinks as a result of a producer's tax. The imposition of a lower bound constraint on each region will counter such an effect on production and the total outputs will be maintained at a certain level. The effects of these arrangements implied in problems A2.5 and A2.6 on the patterns of grain production will be analysed in the following chapters where we will also discuss in more detail why the government imposes these restrictions to achieve its goals.

A3 Data Processing

This chapter will estimate the regional demand functions, production costs and inter-regional transportation costs required to establish the models presented in Chapter A2. It is divided into three sections. Section A3.1 defines the geographical regions and estimates the distances and transportation costs between them. Section A3.2 presents the methodology and estimation of regional demand functions. Section A3.3 estimates production costs, land productivity and land constraints.

A3.1 DEFINITION OF REGIONS AND TRANSPORTATION COSTS

A3.1.1 Regional demarcation

The area under investigation in this study covers the entire mainland of China and Hainan Island, which was established as an independent province in 1989. In China, there are 22 provinces; five autonomous regions; and three metropolitan cities. This study divides the country into 26 demand and supply regions where the demand regions are exactly the same as the supply regions. In principle, each province, or autonomous region, or metropolitan city, forms one separate unit of demand and supply with the following exceptions: (a) Hebei Province, Beijing City and Tianjin City are aggregated into one single unit under the name of 'Beijing'; (b) Yunnan Province and Tibet Autonomous Region are integrated into another single unit called 'Yunnan', partly because they are neighbouring regions and partly because the population and grain production in Tibet are insignificant compared with those of other regions; and (c) Hainan Province has just become independent of Guangdong, but for the purpose of our study, the region 'Guangdong' includes Hainan.

For each region, it is assumed that the production centre is the same as the consumption centre. One city in each region is chosen to be the centre for both demand and supply. The centre is also called the base point. Since the capital cities of most of the provinces and autonomous regions are centrally located, they are taken as the base points for

Table A3.1   Regional definitions

Region Number	Name	Areas covered by regions	Centres
1	Beijing	Beijing Tianjin Hebei	Beijing
2	Shanxi	Shanxi	Taiyuan
3	In.Mon	Inner Mongolia	Huhot
4	Liaoning	Liaoning	Shenyang
5	Jilin	Jilin	Changchun
6	Heilongjiang	Heilongjiang	Haerbin
7	Shanghai	Shanghai	Shanghai
8	Jiangsu	Jiangsu	Zhengjiang
9	Zhejiang	Zhejiang	Jinghua
10	Anhui	Anhui	Hefei
11	Fujian	Fujian	Nanping
12	Jiangxi	Jiangxi	Nanchang
13	Shangdong	Shangdong	Jinan
14	Henan	Henan	Zhengzhou
15	Hubei	Hubei	Wuhan
16	Hunan	Hunan	Changsa
17	Guangdong	Guangdong & Hainan	Guangzhou
18	Guangxi	Guangxi	Nanning
19	Sichuan	Sichuan	Chendu
20	Guizhou	Guizhou	Guiyang
21	Yunnan	Yunnan & Tibet	Kunming
22	Shaanxi	Shaanxi	Xian
23	Gansu	Gansu	Lanzhou
24	Qinghai	Qinghai	Xining
25	Ningxia	Ningxia	Yinchuang
26	Xinjiang	Xinjiang	Uremqi

convenience except Jiangsu, Zhejiang and Fujian, the centres of which
are Zhenjiang, Jinghua and Nanping respectively. Table A3.1 sums up
the above definitions of regions and centres.

A3.1.2   Estimation of transportation costs

(1)   Transportation distances

As the centres for demand and supply in each region are at the same
'point', intra-regional transportation costs are entirely ignored. In

China, because inter-regional transportation of grain mainly depends on the railway system (about 65 per cent), and waterways (about 30 per cent), the distances between each pair of regions are measured using the railway transportation networks existing in 1985. These are presented in Table A3.2.

The distances between those regions located along the Changjiang River are measured along the river rather than railways which are usually longer than the waterways. There is no direct railway linking Guangzhou, the capital of Guangdong, and Nanping, the centre of Fujian, so the distance between these two regions are measured along the South China Sea. The distances between other regions along the coast do not differ much from the distances by rail.

Table A3.2   Transportation distances between regions (km)

j	d1j	d2j	d3j	d4j	d5j	d6j	d7j	d8j	d9j	d10j	d11j	d12j	d13j
1	0	607	649	800	1157	1442	1520	1321	1639	1172	2233	2166	494
2	607	0	621	1408	1755	2049	1957	1758	2076	1609	2713	2280	1102
3	649	621	0	1450	1797	2091	2170	1971	2288	1822	2882	2815	1144
4	800	1408	1450	0	346	641	2321	2122	2440	1973	3034	2967	1295
5	1157	1755	1797	346	0	294	2668	2469	2786	2320	3381	3314	1642
6	1442	2049	2091	641	294	0	2962	2763	3081	2614	3209	3609	1936
7	1520	1957	2170	2321	2668	2962	0	199	366	629	961	833	1026
8	1321	1758	1971	2122	2469	2763	199	0	566	430	1160	1033	826
9	1640	2076	2288	2440	2786	3081	366	566	0	466	594	466	1144
10	1172	1609	1822	1973	2320	2614	629	430	466	0	1061	933	677
11	2233	2713	2882	3034	3381	3209	961	1160	594	1061	0	432	1738
12	2166	2280	2815	2967	3314	3609	833	1033	466	933	432	0	1611
13	494	1102	1144	1295	1642	1936	1026	826	1144	677	1738	1611	0
14	748	862	1362	1549	1895	2190	1096	897	1214	748	1851	1418	748
15	1313	1426	1962	2113	2460	2755	917	1117	917	791	1286	853	1313
16	1736	1850	2386	2537	2884	3178	1264	1295	897	1182	862	430	1736
17	2462	2571	3120	3272	3619	3913	1914	2113	1547	1949	1356	1080	2471
18	2824	2839	3460	3624	3972	4266	2267	2466	1900	2302	1866	1433	2824
19	2128	1520	2142	2929	3275	3570	2260	2460	2260	2133	2170	2195	2112
20	2862	2255	2871	3664	4011	4304	2224	2423	1857	2259	1435	1390	2357
21	3046	2438	3060	3846	4193	4488	2845	3044	2478	2880	2056	2011	3031
22	1293	686	1307	2093	2440	2735	1626	1426	1744	1277	2311	1948	1277
23	1964	1357	1356	2765	3113	3406	2297	2098	2225	1758	2989	2545	1949
24	2261	1654	1653	3062	3409	3703	2594	2395	2404	2055	3279	2842	2246
24	1511	1483	862	2311	2659	2953	3089	2890	2720	2253	3477	3040	2006
26	3957	3646	3644	4757	5104	5399	4586	4386	4217	3750	4974	4537	3941

Continued overleaf

Table A3.2 *continued*

j	d14j	d15j	d16j	d17j	d18j	d19j	d20j	d21j	d22j	d23j	d24j	d25j	d26j
1	748	1313	1736	2462	2824	2128	2793	3046	1293	1964	2261	1511	3957
2	862	1426	1850	2571	2839	1520	2255	2438	686	1357	1654	1483	3646
3	1362	1962	2386	3120	3460	2073	2874	3060	1307	1356	1653	862	3644
4	1549	2113	2537	3272	3624	2929	3664	3846	2093	2765	3062	2311	4757
5	1895	2460	2884	3619	3972	3275	4011	4193	2440	3113	3409	2659	5104
6	2190	2755	3178	3913	4266	3570	4304	4488	2735	3406	3703	2953	5399
7	1096	917	1264	1914	2267	2260	2224	2845	1626	2297	2594	3089	4586
8	897	1117	1295	2113	2466	2460	2423	3044	1426	2098	2395	2890	4386
9	1214	917	897	1547	1900	2950	1857	2478	1744	2225	2404	2720	4217
10	748	791	1182	1949	2302	2133	2259	2880	1277	1758	2055	2253	3750
11	1851	1286	862	1356	1866	2170	1435	2056	2311	2989	3279	3477	4974
12	1418	853	430	1080	1433	2195	1390	2011	1948	2545	2842	3040	4537
13	748	1313	1736	2471	2824	2112	2357	3031	1277	1949	2246	2006	3941
14	0	564	988	1723	2076	1364	2033	2282	529	1202	1498	1696	3193
15	564	0	422	1158	1511	1342	1467	2089	1164	1691	1988	2187	3684
16	988	422	0	735	1006	1765	1044	1666	1517	2115	2411	2611	4108
17	1723	1158	735	0	804	2415	1680	2301	2253	3273	3570	3769	5266
18	2076	1511	1006	804	0	1539	804	1425	2374	2663	2960	3158	4655
19	1364	1342	1765	2415	1539	0	735	917	834	1124	1420	1619	3116
20	2033	1467	1044	1680	804	735	0	621	1569	1859	2155	2354	3851
21	2282	2089	1666	2301	1425	917	621	0	1753	2042	2339	2537	4034
22	529	1095	1517	2253	2374	834	1569	1753	0	640	936	1135	2632
23	1202	1691	2115	3273	2663	1124	1859	2042	640	0	296	494	1991
24	1498	1988	2411	3570	2960	1420	2155	2339	936	296	0	791	2288
24	1696	2187	2611	3769	3158	1619	2354	2537	1135	494	791	0	2486
26	3193	3684	4108	5266	4655	3116	3851	4034	2632	1991	2288	2486	0

(2) Transportation costs

Per unit transportation costs between any pair of regions are assumed to be independent of the volume of flows, but dependent upon the distances. Suppose the distances between region i and region j are denoted as d_{ij} (for all i and j), the unit transportation costs between i and j are denoted as t_{ij} (for all i and j), then according to the assumption made above, t_{ij} is the function of d_{ij}, or

$$t_{ij} = \alpha + \beta d_{ij} \qquad \text{for all } i \text{ and } j \qquad (A3.1)$$

where α and β are constants

Table A3.3   Grain transshipment costs by railways (ton/¥)

Distance (km)	Costs (¥/ton)	Distance (km)	Costs (¥/ton)
100	2.00	1925	23.30
145	2.60	2050	24.60
205	3.40	2250	27.00
265	4.40	2450	29.40
325	5.40	2650	31.80
385	6.30	2850	34.20
460	7.50	3050	36.60
540	8.60	3250	39.00
620	9.60	3450	41.40
700	10.60	3650	43.80
780	11.50	3850	46.20
860	12.50	4050	48.60
940	13.40	4250	51.00
1025	14.40	4450	53.40
1125	15.40	4650	55.80
1225	16.40	4850	58.20
1325	17.40	5050	60.60
1425	18.40	5250	63.00
1525	19.40	5450	65.40
1625	20.40	5650	67.80
1725	21.30	5850	70.20
1825	22.30	—	—

Source: Ministry of Railways (1985).

The values of α and β can be estimated with the data given in Table A3.3. The data was collected from a handbook published by the State Railway Ministry which set the standard prices of grain transportation by railways. The prices do not include the costs of loading, unloading, waste and other expenses.

In China, unit transportation costs depend on the means of transportation. The unit transportation costs by waterways are slightly lower than those by railways and the costs by highways are about five or six times as high as those by railways. Transshipments between regions along the Changjiang river and the coastal areas are more likely to be carried out through waterways. However, even the transshipment between these regions has to rely on the highways (trucks) and the railways to a large extent. This tends to make the estimations of the combined (average) transshipment costs much more difficult.

To simplify the estimation, we employ the transshipment costs based on railway transportation for all the regions but some modifications are made for the costs between those regions located along the river and the coast. Therefore, the unit transportation costs between those regions are slightly lower (assumed to be 15 per cent lower) than they should be if the transhipment were entirely carried out via railways.

To estimate equation (A3.1), the costs in Table A3.3 are increased by one-and-a-half times to include the costs of loading, unloading, storage, waste, higher costs by truck and other expenses. The estimated results are reported as follows:

$$t_{ij} = 0.42 + 0.0029 \times d_{ij} \qquad (\text{¢/kg–km})$$
$$(0.0349) \quad (0.0003)$$

$$R^2 = 0.9803 \tag{A3.2}$$

The values in the brackets are standard errors. Both α and β are significantly different from zero and the R^2 is very high. The unit transportation costs presented in Table A3.4 are computed by using this equation and the data in Table A3.2, and with some modifications mentioned above for the coastal and riverside regions.

A3.2 REGIONAL DEMAND FUNCTIONS

Total regional demands may be obtained by multiplying per capita demands with total regional populations. Therefore,

$$d_i = TP_i \times y_i \qquad \text{for all } i \tag{A3.3}$$

Where d_i, TP_i, y_i are respectively total demand, population and per capita demand in region i.

For a particular year, TP_i can be obtained from the population census. Average per capita demand, however, is determined by many factors, such as price and income. For simplicity, it is assumed that y_i is linearly determined by grain price and per capita expenditure, or,

$$y_i = \beta_{0i} - \beta_{1i}P_i + \beta_{2i}I_i + U_i \qquad \text{for all } i \tag{A3.4}$$

where β_{0i}, β_{1i} and β_{2i} are coefficients and U_i is a disturbance term.

Table A3.4 Inter-regional transportation costs of grain in 1985 (¢/100kg)

j	t1j	t2j	t3j	t4j	t5j	t6j	t7j	t8j	t9j	t10j	t11j	t12j	t13j
1	0	218	230	274	377	460	411	425	517	382	587	670	185
2	218	0	222	450	551	636	609	552	644	508	828	703	361
3	230	222	0	462	563	648	671	613	705	570	878	858	373
4	274	450	462	0	142	228	608	657	749	614	784	902	355
5	377	551	563	142	0	127	815	758	850	714	1022	1003	518
6	460	636	648	228	127	0	901	843	935	800	972	1088	603
7	412	609	671	608	815	901	0	99	148	191	273	283	339
8	425	552	613	657	758	843	99	0	206	166	378	341	281
9	517	644	705	749	850	935	148	206	0	177	214	177	373
10	382	508	570	614	714	800	191	166	177	0	349	312	238
11	587	828	878	784	1022	972	273	378	214	349	0	167	464
12	670	703	858	902	1003	1088	283	341	177	312	167	0	509
13	185	361	373	355	518	603	339	281	373	238	464	509	0
14	259	292	437	491	591	677	360	302	394	259	578	453	259
15	422	455	611	655	755	841	262	366	308	231	415	289	422
16	545	578	734	777	878	963	408	417	302	385	292	166	545
17	643	787	947	842	1091	1177	507	557	417	607	370	355	645
18	861	865	1045	1093	1194	1279	699	757	593	709	583	457	861
19	659	483	663	891	992	1077	593	755	697	562	671	678	654
20	872	696	874	1104	1205	1290	687	744	580	697	458	445	725
21	925	749	929	1157	1258	1343	867	925	760	877	638	625	921
22	417	241	421	649	749	835	513	455	548	412	712	607	412
23	611	435	435	844	944	1030	708	650	687	552	909	780	607
24	697	521	521	930	1030	1116	794	736	739	638	993	866	693
24	480	472	292	712	813	898	938	880	830	695	1050	923	623
26	1189	1099	1099	1421	1522	1607	1372	1314	1265	1129	1484	1357	1185

j	t14j	t15j	t16j	t17j	t18j	t19j	t20j	t21j	t22j	t23j	t24j	t25j	t26j
1	259	422	545	643	861	659	852	925	417	611	697	480	1189
2	292	455	578	787	865	483	696	749	241	435	521	472	1099
3	437	611	734	947	1045	643	875	929	421	435	521	292	1099
4	491	655	777	842	1093	891	1104	1157	649	844	930	712	1421
5	591	755	878	1091	1194	992	1205	1258	749	944	1030	813	1522
6	677	841	963	1177	1279	1077	1290	1343	835	1030	1116	898	1607
7	360	262	408	507	699	593	687	867	513	708	794	938	1372
8	302	366	417	557	757	755	744	925	455	650	736	880	1314
9	394	308	302	417	593	897	580	760	548	687	739	830	1265
10	259	231	385	607	709	562	697	877	412	552	638	695	1129
11	578	415	292	370	583	671	458	638	712	909	993	1050	1484
12	453	289	166	355	457	678	445	625	607	780	866	923	1357
13	259	422	545	645	861	654	725	921	412	607	693	623	1185
14	0	205	328	541	644	437	631	704	195	390	476	534	968
15	205	0	164	378	480	366	467	647	379	532	618	676	1110
16	328	164	0	255	334	554	345	525	482	655	741	799	1233
17	541	378	255	0	275	742	529	709	695	991	1077	1135	1569
18	644	480	334	275	0	488	275	455	730	814	900	958	1392
19	437	366	554	742	488	0	255	308	284	368	454	511	945
20	631	467	345	529	275	255	0	222	497	581	667	724	1159
21	704	647	525	709	455	308	222	0	550	634	720	777	1212
22	195	359	482	695	730	284	497	550	0	227	313	371	805
23	390	532	655	991	814	368	581	634	227	0	128	185	619
24	476	618	741	1077	900	454	667	720	313	128	0	271	705
24	534	676	799	1135	958	511	724	777	371	185	271	0	763
26	968	1110	1233	1569	1392	945	1159	1212	805	619	705	763	0

A3.2.1 National demand functions of selected products

Let y_t^k denote national per capita demand for commodity k
($k = 1, 2, \ldots, m$) in year t, and suppose that per capita expenditure
and prices of substitutable commodities are the only explanatory
variables, then the demand function of a particular commodity k can
be represented by equation (A3.5) as follows:

$$y_t^k = f(E_t, P_t^k, P_t^j, U_t) \tag{A3.5}$$

Where E_t, P_t^k, P_t^j are respectively per capita expenditure, the price of
commodity k and the price of competitive commodities of k in year t.
U_t is a disturbance term. The demand functions are estimated in two
different forms: simple linear and double-log.

For a linear function:

$$y_t^k = \beta_0 + \beta_1 E_t - \beta_2 P_t^k + \beta_3 P_t^j + U_t \tag{A3.6}$$

For a double-log function:

$$\ln y_t^k = \varepsilon_0 + \varepsilon_1 \ln E_t - \varepsilon_2 \ln P_t^k + \varepsilon_3 \ln P_t^j + V_t \tag{A3.7}$$

For comparison, the demand functions of three important agricultural
products are estimated for the above two functional forms by using the
data in Table A3.5. Time series data were available from 1952 to 1985.
All functions are well explained by own prices and per capita
expenditure. The effects of competitive prices are insignificant. Therefore, the demand functions reported here have only two explanatory
variables, i.e. own price and per capita expenditure. The effects of
inflation are excluded by deflating the prices and expenditure using the
retailed price index of farm products (Table A3.5).

The estimated results are reported below for each individual
commodity, i.e. grain, meat and edible vegetable oil.

(i) Grain per capita demand function

Double-log function

$$LDG_t = 4.3691 + 0.5032 \times LAE_t - 0.5075 \times LPG_t$$
$$\quad\quad (26.463) \quad\quad (7.179) \quad\quad\quad (6.870)$$
$$R^2 = 0.9566 \quad\quad DW = 2.3140$$

Table A3.5   Per capita consumption of different agricultural products, expenditure and prices, 1970–85

Year	Export per cap. (¥)	Consump. (kg/p.c.)			Prices (¢/kg)			Price index of farm crops
		Meat	Veg. oil	Grain	Grain	Meat	Veg. oil	
1970	140	6.83	1.60	239.69	26.00	116.00	164.00	195.10
1971	142	7.82	1.64	241.39	26.00	116.00	164.00	198.30
1972	147	8.33	1.66	274.55	27.76	116.00	164.00	201.10
1973	155	8.39	1.65	278.44	27.76	116.00	163.82	202.80
1974	155	8.40	1.71	296.56	28.72	120.60	164.34	204.50
1975	158	8.34	1.72	281.56	28.80	119.20	164.08	208.70
1976	161	8.04	1.59	331.35	28.82	115.40	165.30	209.70
1977	165	7.95	1.55	296.21	29.20	117.20	165.30	209.20
1978	175	8.42	1.59	299.82	29.48	117.40	164.66	217.40
1979	197	10.48	1.95	278.91	29.86	133.80	168.96	265.50
1980	227	11.98	2.30	268.44	30.75	164.80	171.35	284.40
1981	249	11.92	2.94	313.82	33.71	170.20	196.74	301.20
1982	267	12.78	3.53	229.35	34.03	171.00	196.91	307.80
1983	290	13.45	4.02	243.02	35.14	178.20	192.98	321.30
1984	330	14.27	4.70	293.76	35.85	221.00	207.13	334.40
1985	407	15.31	5.13	318.13	38.33	286.10	217.92	362.90

Notes:
Meat consumption includes the consumption of pork, beef and mutton. The mixed retail price of meat is the weighted averages of these three types of meats.
Sources: The consumptions between 1970 and 1983 are from SSB, *CSYB* (1984), pp. 440–1; data for 1984 and 1985 are from SSB, *CSYB* (1985), pp. 543–9, and 1986, pp. 551–2. The price indices between 1970 and 1983 are from SSB, *CSYB* (1984), pp. 425; 1984 from SSB, *CSYB* (1985), pp. 529; 1985 from SSB, *CSYB* (1986), pp. 535. Data for per capita living expenditure is from SSB, *CSYB* (1986), pp. 556.

Linear function

$$DG_t = 207.83 + 127.4430 \times AE_t - 846.6579 \times PG_t$$
$$(8.220) \qquad (7.119) \qquad\qquad (8.184)$$
$$R^2 = 0.9605 \qquad DW = 2.2036$$

(ii) Meat per capita demand function

Double-log function

$$LDM_t = 2.0000 + 1.4669 \times LAE_t - 1.9098 \times LPMt$$
$$ (14.680) \qquad (8.016) \qquad\qquad (5.122)$$
$$R^2 = 0.8970 \qquad DW = 1.162$$

Linear function

$$DM_t = 13.9500 + 18.5854 \times AE_t - 25.8645 \times PM_t$$
$$ (3.136) \qquad (9.255) \qquad (5.151)$$
$$R^2 = 0.9100 \qquad DW = 1.0643$$

(iii) Vegetable oil per capita demand function

Only the double-log demand function is used for vegetable oil demand:

$$LDO_t = 0.8310 + 2.1551 \times LAE_t - 1.2863 \times LPO_t$$
$$ (3.803) \qquad (4.814) \qquad\qquad (3.137)$$
$$R^2 = 0.8869 \qquad DW = 0.7343$$

where, LDG_t, LDM_t and LDO_t are respectively per capita demands for grain, meat and vegetable oil in logarithmic forms in year t; DG_t, DM_t and DO_t are respectively per capita demands for grain, meat and vegetable oil in year t; AE_t and LAE_t are per capita expenditure (deflated) and its logarithmic form in year t; PG_t, PM_t and PO_t are respectively the average retailed prices (deflated) of grain, meat and vegetable oil in year t; LPG_t, LPM_t and LPO_t are respectively the logarithmic forms of PG_t, PM_t and PO_t.

Figures in the parentheses are t-values. All the t-values of these demand functions are very high, so are the R^2s. All the three demand functions are well explained by prices and expenditure. The price and expenditure elasticities are reported in Table A3.6.

It is obvious that the demands for meat and vegetable oil are much more elastic than the demand for grain, with respect to either price or expenditure. In fact, according to our estimated results, meat and vegetable oil are both luxury goods in China while grain is a necessity.

Table A3.6   Price and expenditure elasticities of demands for grain, meat and vegetable oil, 1985

Function forms	Depend. variable	Mean value	Price elasticity		Expenditure elasticity	
			Symbol	Values	Symbol	Values
Double-log	DG	—	E_{GP}	−0.5075	E_{GE}	0.5032
	DM	—	E_{MP}	−1.9098	E_{ME}	1.4669
	DO	—	E_{OP}	−1.2863	E_{OE}	2.1551
Linear	DG	206.18	E_{GP}	−0.5121	E_{GE}	0.5041
	DM	10.17	E_{MP}	−1.8622	E_{ME}	1.4903

Note: $P_G = 0.1247$ and $P_M = 0.7324$ are average values.

Furthermore, the price and expenditure elasticities are very close for both linear and double-log demand functions. Therefore, it is difficult to tell which one of these two functional forms is more appropriate for the demands of grain and meat in China. For convenience, however, the elasticities of grain demand used in the following analysis will be the price and expenditure elasticities derived from the linear demand function.

A3.2.2   Per capita expenditures in 1985

Table A3.7 presents the per capita expenditures for all the regions.

A3.2.3   Regional grain consumption in 1985

Data for regional grain consumption is available only for the rural population, but not for the urban residents. Therefore, regional consumptions of grain for the urban population must be estimated before national average grain consumptions are calculated.

(1)   For the urban population

Suppose per capita grain consumption of the urban population in 1985 is expressed by the following equation:

$$UG_i = b0 + b1 \times UE_i - b2 \times PG_i \qquad \text{for all } i \qquad (A3.8)$$

where UG_i is urban per capita grain consumption in 1985 in region *i*.

Table A3.7   Regional per capita expenditures in 1985 (¥/p.c)

Region	Average	Rural	Urban
1	458	371	722
2	354	261	713
3	419	325	634
4	581	405	833
5	512	377	750
6	502	328	763
7	961	850	1024
8	466	410	756
9	451	398	715
10	337	289	648
11	419	358	717
12	366	326	532
13	376	334	658
14	293	240	710
15	406	327	720
16	370	326	646
17	493	383	871
18	311	270	581
19	338	287	652
20	288	232	698
21	313	257	699
22	345	267	700
23	299	210	780
24	444	299	812
25	370	286	670
26	484	336	658
National	401	320	730

Notes: All the expenditures are calculated with current prices.
Source: SSB, *CSYB* (1987), People's Livings.

UE_i is urban per capita expenditure in region i. PG_i is the average retailed price of grain in region i. $b0$, $b1$ and $b2$ are coefficients. Because it is impossible to estimate $b0$, $b1$ and $b2$ from available data for each individual region, before the regional per capita grain consumptions are estimated for the urban population, two further assumptions are needed: (a) different regions have the same coefficients $b0$, $b1$ and b2; and (b) the values of $b0$, $b1$ and b2 are determined by the elasticities in Table A3.6, urban per capita expenditure and the average retailed price of grain in 1985. Thus,

$$b1 = E_{GE} \times (AUG_{85}/AUE_{85})$$
$$b2 = -E_{GP} \times (AUG_{85}/APG_{85})$$
$$b0 = AUG_{85} + b2 \times APG_{85} - b1 \times AUE_{85}$$

where AUG_{85} is urban per capita grain consumption in 1985. AUE_{85} is urban per capita expenditure in 1985. APG_{85} is the national average price of grain in 1985. $E_{GE} = 0.5041$ (Table A3.6) is the expenditure elasticity of grain demand. $E_{GP} = 0.5121$ (Table A3.6) is the price elasticity of grain demand. $AUG_{85} = 254.35$ kg (Table A3.5). $APG_{85} = 50.05$ (¢/kg) (Table A3.10, APP1). $AUE_{85} = 730$ (¥/p.c.) (Table A3.7).

The values of APG_{85} should be equal to 38.33 ¢/kg according to Table A3.5, but since some adjustments for grain prices will be made in Table A3.10, the adjusted average price 50.05 ¢/kg is used here. The reason for this will be given below.

Given the above values, the coefficients are estimated as follows:

$$b1 = 0.5041 \times (254.35/730) = 0.1756$$
$$b2 = 0.5121 \times (254.35/50.05) = 2.6025$$
$$b0 = 254.35 + 2.6364 \times 50.05 - 0.1756 \times 730 = 256.42$$

Thus,

$$UG_i = 256.42 + 0.1756 \times UE_i - 2.6025 \times PG_i, \text{ for all } i \qquad (A3.9)$$

Because $b0$, $b1$ and $b2$ are the same for all the regions, the regional differences of estimated grain consumption will be explained by price and expenditure. In (A3.9), UE_i are from Table A3.7, PG_i are $APP1$ from Table A3.10. Let us take Beijing as an example: per capita expenditure in Beijing in 1985 was 722 yuan and the grain price was 53.37¢/kg, therefore, per capita grain consumption in Beijing was:

$$256.38 + 0.1756 \times 722 - 53.37 \times 2.6025 = 244.31 \quad \text{(kg/p.c.)}$$

Estimated results for all regions are presented in Table A3.8.

(2) For the whole population

Regional per capita grain consumption for all the urban and rural people is the weighted average of regional consumptions by the urban and rural people with the respective populations as weights. Let DG_i denote average per capita grain consumption in region i for all the people, thus

$$DG_i = 1/TP_i \times (UP_i \times UG_i + RP_i \times RG_i) \qquad \text{(A3.10)}$$

where TP_i is total population in region i;
         UP_i is urban population in region i;
         UG_i is per capita grain consumption for the urban population
         in region i.
         RP_i is rural population in region i;
         RG_i is per capita grain consumption for the rural population in
         region i.

The estimated results are shown in the last column of Table A3.8.

Table A3.8   Regional population and estimated per capita grain consumption
               in 1985

j	Population ('000)			Grain consumption (kg/p.c.)		
	Total	*Rural*	*Urban*	*Rural*	*Urban*	*Average*
1	73162	55384	17778	215.34	244.31	222.38
2	26265	20990	5275	223.15	249.05	228.35
3	20067	14346	5720	266.22	251.84	262.11
4	36862	21837	15025	242.66	288.13	261.19
5	22980	14610	8370	286.92	281.00	284.76
6	33114	19753	13361	258.63	269.44	262.99
7	12167	3733	8434	263.84	300.33	289.14
8	62135	48043	14092	291.23	251.37	282.19
9	40296	32980	7316	298.62	246.25	289.11
10	51558	44316	7242	276.69	231.76	270.38
11	27131	22659	4472	261.76	246.87	259.30
12	34602	28355	6248	325.20	214.14	305.16
13	76947	66776	10171	221.12	233.33	222.73
14	77128	68115	9014	236.85	236.87	236.85
15	49309	38825	10484	294.89	249.16	285.17
16	56227	48260	7967	326.06	250.27	315.32
17	62531	49432	13099	256.11	291.92	263.61
18	38729	34018	4711	253.40	245.23	252.41
19	101875	87043	14832	252.98	247.92	252.24
20	29679	26090	3589	219.11	263.88	224.52
21	36057	31850	4206	224.51	258.75	228.50
22	30017	24620	5397	256.50	243.93	254.24
23	20413	17299	3114	244.18	252.44	245.44
24	4074	2912	1161	226.14	256.31	234.68
25	4146	3313	832	274.25	230.88	265.48
26	13611	9225	4387	216.58	221.70	218.24
Sum	1041084	834786	206297	258.30	255.14	257.76

Source: Rural per capita grain consumption SSB, CAYB (1986), p. 588.

Data Processing 261

Table A3.9   Regional grain outputs and composition in 1985

j	Output (10,000 ton)				Share of total (%)		
	Total	Rice	Wheat	Other grains	Rice	Wheat	Other
1	2326.8	120.3	860.9	1345.6	5.17	37.00	57.83
2	822.7	5.8	295.1	521.8	0.70	35.87	63.43
3	604.1	7.8	148.5	447.8	1.29	24.58	74.13
4	976.0	263.0	2.8	710.2	26.95	0.29	72.77
5	1225.3	184.1	10.3	1030.9	15.02	0.84	84.13
6	1330.0	162.9	276.8	890.3	12.25	20.81	66.94
7	213.8	153.9	21.8	38.1	71.98	10.20	17.82
8	1621.3	1357.8	90.4	173.1	83.75	5.58	10.68
9	3126.5	1638.5	829.4	658.6	52.41	26.53	21.07
10	2168.0	1162.9	605.9	399.2	53.64	27.95	18.41
11	794.4	681.1	17.4	95.9	85.74	2.19	12.07
12	1533.5	1475.8	10.4	47.3	96.24	0.68	3.08
13	3137.7	62.5	1496.1	1579.1	1.99	47.68	50.33
14	2710.5	226.3	1528.2	956.0	8.35	56.38	35.27
15	2216.1	1571.7	345.4	299.0	70.92	15.59	13.49
16	2514.3	2338.8	29.2	146.3	93.02	1.16	5.82
17	1737.9	1561.4	7.9	168.6	89.84	0.45	9.70
18	1117.1	986.0	1.0	130.1	88.26	0.09	11.65
19	3830.7	1926.1	625.6	1279.0	50.28	16.33	33.39
20	595.0	324.3	29.5	241.2	54.50	4.96	40.54
21	988.1	483.3	73.7	431.1	48.91	7.46	43.63
22	951.9	88.3	423.3	440.3	9.28	44.47	46.25
23	530.5	1.8	314.8	213.9	0.34	59.34	40.32
24	100.3	0.0	63.0	37.3	0.00	62.81	37.19
25	139.5	41.9	58.6	39.0	30.04	42.01	27.96
26	498.8	30.6	314.5	153.7	6.13	63.05	30.81
Sum	37910.8	16856.9	8580.5	12473.4	44.46	22.64	32.94

Source: The first four columns are from SSB, *CAYB* (1986) Agriculture. The last three columns are derived from the first four.

A3.2.4   Regional grain demand prices in 1985

The State Price Bureau published a handbook at the end of 1984, which contains the average procurement prices (APP) for 1985 of different grain crops, including rice, wheat, corn and other grains. The average price means the 'mixed' average of the basic procurement price and the premium price, with the latter being 50 per cent higher than the

Table A3.10   Regional estimated prices in 1985 (¢/kg)

j	Rice	Wheat	Other	APP	APP₁	APP₂	APP₃
1	46.32	46.80	31.84	38.12	53.37	47.65	64.81
2	42.66	45.64	31.08	36.38	50.94	45.48	61.85
3	42.80	44.42	27.44	31.81	44.54	39.77	54.08
4	42.66	44.28	27.24	31.44	44.02	39.31	53.45
5	44.80	44.28	26.50	29.40	41.16	36.75	49.98
6	46.10	45.10	25.80	32.30	45.22	40.38	54.92
7	38.08	41.08	32.00	37.30	52.22	46.63	63.41
8	38.00	42.40	31.60	37.56	52.59	46.95	63.86
9	37.66	42.40	31.32	37.58	52.61	46.98	63.89
10	37.80	42.60	31.60	38.00	53.20	47.50	64.60
11	37.80	42.40	31.80	37.18	52.05	46.47	63.20
12	37.40	42.20	31.32	37.25	52.14	46.56	63.32
13	38.40	45.26	31.20	38.05	53.27	47.56	64.68
14	37.30	45.10	31.32	39.59	55.42	49.49	67.30
15	36.40	42.66	31.32	36.69	51.37	45.86	62.37
16	32.80	42.20	31.32	32.82	45.95	41.03	55.80
17	32.26	42.50	31.50	32.23	45.13	40.29	54.80
18	31.18	41.50	30.16	31.07	43.50	38.84	52.82
19	32.90	40.00	32.00	33.76	47.26	42.20	57.39
20	32.20	38.00	30.00	31.60	44.23	39.49	53.71
21	33.20	42.20	31.32	33.05	46.27	41.31	56.19
22	31.18	44.82	31.00	37.16	52.03	46.45	63.18
23	31.18	44.28	30.52	38.69	54.16	48.36	65.77
24	31.18	44.28	30.52	39.16	54.83	48.95	66.58
25	40.50	44.28	30.52	39.30	55.02	49.12	66.81
26	40.36	46.58	30.52	41.25	57.75	51.56	70.12
Sum	—	—	—	35.75	50.05	44.69	60.75

Notes:
$APP = RIP \times RIR + WHP \times WHR + OTP \times OTR$; $APP_1 = APP \times 1.4$;
$APP_2 = APP \times 1.25$; $APP_3 = APP \times 1.7$. RIR, WHR, OTR are the shares of
the outputs of rice, wheat and other grains to the total output of grains from
Table A3.9. RIP, WHP and OTP are respectively the prices of rice, wheat and
other grains shown above.
Source: Prices in the first three columns are collected from a handbook of the
SPB in Beijing.

former. The basic procurement price was usually much lower than the
open market price. It used to be about 20 yuan/100kg before 1978 for
rice and wheat but has increased two or three times since 1979. In 1985
the basic procurement price for rice, for example, was on average about
30 yuan/100kg. In 1985, a mixed procurement price of grain for each

province was introduced to replace the former two different procurement prices existing between 1978 and 1984 (see detailed discussion in Chapter 4). The mixed procurement price is a weighted average of the former basic and premium prices. The weights attached to the basic and premium prices were different for different regions. For instance, the basic procurement price of unprocessed rice in Jilin (region 5) was set at 32.00 yuan/100kg. The premium price was 50 per cent higher or 48 yuan/100kg. Twenty per cent of total procurement would be paid at the basic price and the rest (80 per cent) would be paid at the premium price. Thus, the 'mixed' average procurement price of rice in Jilin was fixed at 44.80 yuan/100kg (Table A3.10, column 1).

In order to calculate the average prices of all kinds of cereal products, grain is divided into three groups, i.e., rice, wheat and other grains. The total outputs, the outputs by crop group and their shares in total outputs for all regions are presented in Table A3.9.

The APPs for these three crops in each region are presented in the first three columns of Table A3.10. The APP on column 4 of the table are the weighted average prices of all the crops with the outputs of these three groups of grain as the weights.

A3.2.5 Regional demand equations

Based on the estimates of regional per capita expenditures, grain consumptions and grain prices in the previous sections, regional grain demand equations to be used for the models in Chapter A2 are estimated in this section.

(1) Regional demand equations: quantity formulations

Let us restate equation (A3.4) here by dropping the disturbance term,

$$y_i = \beta_{0i} - \beta_{1i} P_i + \beta_{2i} I_i, \text{ for all } i \qquad (A3.11)$$

Where y_i, P_i and I_i are respectively the regional per capita demand for grain, price of grain and per capita expenditure in region i. β_{0i}, β_{1i} and β_{2i} are coefficients. Before defining the regional demand equations, three more assumptions must be made: (i) β_{1i} and β_{2i} are the same for all the regions; (ii) the intercept β_{0i} for all i may be different and expenditure I_i is exogenous (and therefore fixed) in (A3.11); and (iii) the actual prices (market demand prices) in all regions are assumed to be 40 per cent higher than the average procurement prices (*APP*).[1]

Therefore, the values of APP1 in Table A3.10 are used in the following estimation.

Equation (A3.11) can be rewritten as,

$$y_i = \Theta_i - \beta_1 P_i \qquad \text{for all } i \qquad (A3.12)$$

Where $\quad \Theta_i = \beta_{0i} + \beta_{2i} I_i \qquad \text{for all } i \qquad (A3.13)$

β_1 can be readily estimated by the following equation,

$$\beta_1 = -E_{GP} \times (ADG_{85}/APG_{85}) \qquad (A3.14)$$

The values of Θ_i are still unknown, but they can be estimated by equation (A3.15).

$$\Theta_i = DG_i + \beta_1 P_i \qquad (A3.15)$$

Where: DG_i = estimated per capita grain consumption in 1985 in region i (Table A3.8, last column).

ADG_{85} = 257.67 (kg/p.c.; Table A3.8, last column) is the national average grain consumption per capita in 1985.

APG_{85} = 50.05 (¢/kg) (Table A3.10, APP1), is the national average estimated grain demand price in 1985.

P_i = APP1 in Table A3.10, for all i, are the estimated regional demand prices of grain in 1985.

E_{GP} = −0.5121 (Table A3.6) is the estimated price elasticity of grain demand in 1985.

Thus,

$$\beta_1 = 0.5121 \times (257.67/50.05) = 2.6364 \quad \text{and}$$

$$y_i = \Theta_i - 2.6264 \times P_i \qquad (A3.16)$$

The values of Θ_i vary by region: for instance, the intercept for region 'Beijing' is

$$\Theta_1 = DG_1 + 2.6364 \times P_1 = 222.38 + 2.6364 \times 53.37$$
$$= 363.09$$

Therefore, the demand equation for Beijing is

$$y_i = 363.09 - 2.6364 \times P_i$$

Table A3.11 Demand equation coefficients, 1985

j	Θ	β_I	A_0 ('000t)	A_1 ('000t)	ETGD ('000t)	h_0 ('000t)	h_1 ('000t)
1	363.09	2.6364	26564.54	192.884	16269.69	137.723	0.0051845
2	362.64	2.6364	9524.86	69.245	5997.66	137.553	0.0144415
3	379.52	2.6364	7615.92	52.905	5259.69	143.956	0.0189019
4	377.25	2.6364	13906.31	97.183	9628.16	143.094	0.0102899
5	393.27	2.6364	9037.45	60.584	6543.88	149.171	0.0165059
6	382.22	2.6364	12656.84	87.302	8708.68	144.978	0.0114545
7	426.82	2.6364	5193.08	32.077	3517.91	161.894	0.0311749
8	420.83	2.6364	26148.27	163.813	17533.88	159.623	0.0061045
9	427.83	2.6364	17239.66	106.236	11650.07	162.276	0.0094130
10	410.63	2.6364	21171.47	135.927	13940.16	155.756	0.0073569
11	396.52	2.6364	10758.02	71.528	7035.20	150.403	0.0139805
12	442.63	2.6364	15315.75	91.225	10559.02	167.890	0.0109619
13	363.17	2.6364	27944.48	202.863	17138.70	137.750	0.0049294
14	382.97	2.6364	29538.05	203.340	18268.13	145.264	0.0049179
15	420.59	2.6364	20738.84	129.998	14061.30	159.532	0.0076924
16	436.47	2.6364	24541.39	148.237	17729.57	165.555	0.0067460
17	382.58	2.6364	23923.15	164.857	16483.84	145.115	0.0060659
18	367.09	2.6364	14216.89	102.105	9775.46	139.238	0.0097938
19	376.85	2.6364	38391.18	268.583	25697.23	142.940	0.0037232
20	341.14	2.6364	10124.76	78.246	6663.64	129.397	0.0127803
21	350.49	2.6364	12637.53	95.061	8238.93	132.942	0.0105196
22	391.40	2.6364	11748.79	79.137	7631.53	148.462	0.0126363
23	388.23	2.6364	7925.01	53.817	5010.16	147.259	0.0185815
24	379.23	2.6364	1544.99	10.741	956.10	143.844	0.0931039
25	410.53	2.6364	1702.05	10.931	1100.68	155.715	0.0914870
26	370.50	2.6364	5042.82	35.884	2970.53	140.531	0.0278675

Notes:
$\Theta_i = DG_i + 2.6364 \times APP_{1i}$; $A_{0i} = (1/1000) \times (\Theta_i \times TP_i)$;
$A_{1i} = (1/1000) \times (2.6364 \times TP_i)$; $ETGD_i = A_{0i} - A_{1i} \times APP_{1i}$,
$h_{0i} = A_{0i}/A_{1i}$; $h_{1i} = 1/A_{1i}$.

Where TP_i and APP_{1i} are respectively the total population and estimated demand price of grain in region i.

All the values of Θ_i for all i are given in the first column in Table A3.11. The demand equation for the whole population can be thought of as equation (A3.16) multiplied by regional total populations. Thus

$$d_i = TP_i \times y_i = TP_i \times \Theta_i - 2.6364 \times TP_i \times P_i$$
$$\text{or } d_i = A_{0i} - A_{1i} \times P_i \text{ for all } i \qquad (A3.17)$$

$$A_{0i} = TP_i \times \Theta_i$$

$$A_{1i} = 2.6364 \times TP_i$$

For instance, the demand equation for the whole population in region Beijing is:

$$d_i = (363.09 \times 73.162) - (2.6364 \times 73.162 \times P_i)$$

or $\qquad d_i = 26564.39 - 192.884 \times P_i)$ ('000 ton)

All the values of A_{0i} and A_{1i} are given in Table A3.11.

(2) Regional demand equations: price formulations

The demand equations in the models of Chapter A2 are all in the price form:

$$P_i = h_{0i} - h_{1i} \times d_i \qquad \text{for all } i \tag{A3.18}$$

By inversion of (A3.18),

$$h_{0i} = A_{0i}/A_{1i} \qquad \text{for all } i \tag{A3.19}$$
$$h_{1i} = 1/A_{1i} \qquad \text{for all } i \tag{A3.20}$$

For instance, the demand equation for region Beijing in price form is given as,

$$P_i = 137.7183 - 0.0051845 \times d_i$$

The values of h_{0i} and h_{1i} for all the regions are presented in the last two columns of Table A3.11. The data in Table A3.11 are used to run the models in Chapter A2.

A3.2.6 Local fixed demands

The estimations of regional grain demands presented from sections A2.1 to A2.6 are for human consumption only (or demands). The demands by animals, seeds, local storage and waste are not included. To simplify the programming, all these demands are treated as local fixed and necessary demands. This means that they are not tradable.

Let d_i^0 denote the fixed demand in region i, it will then be the function of the numbers of animals, livestock products, seed and waste, etc.

Before d_i^0 are estimated, it is assumed that: (i) For each unit of animal or animal product, the requirement of grain is a fixed constant for all the regions except Qinghai, Inner Mongolia and Xinjiang where animals rely less on grain than those in other regions; and (ii) The requirement for seeds and waste is assumed to be 3.6 per cent of the regional total outputs for every region.

The outputs of meat (pork, beef and mutton), milk, eggs and the numbers of draft animals in all regions are presented in Table A3.12. Assuming the production of each Kilogramme of meat requires 3kg of grain, each kg of milk requires 0.7kg of grain, each kilo of eggs requires 2.8kg of grain, and each draft animal requires 210kg of grain each year. Then,

$$d_i^0 = MO_i \times 3.0 + MLO_i \times 0.7 + EO_i \times 2.8$$
$$+ NA_i \times 0.21 + 0.036 \times TO_i \qquad \text{for all } i \qquad (A3.22)$$

For the three pastoral regions, the local demands are estimated as follows:

Qinghai: $d^0 = MO \times 3.0 + MLO \times 0.7 + EO \times 2.8 + NA \times 0.08 = 24$
In.Mon: $d^0 = MO \times 3.0 + MLO \times 0.7 + EO \times 2.8 + NA \times 0.12 = 3$
Xinjiang: $d^0 = MO \times 3.0 + MLO \times 0.7 + EO \times 2.8 + NA \times 0.12 = 26$

where MO_i is output of meat in region i
MLO_i is output of milk in region i
EO_i is output of egg in region i
NA_i is number of draft animals in region i
TO_i is total grain output in region i (Table A3.9)

Let us take Beijing as an example,

$$d_i^0 = 1085 \times 3.0 + 281 \times 0.7 + 581 \times 2.8 + 4965 \times 0.21 + 0.036 \times 23268$$
$$= 6958.80 \qquad (\text{'000 ton})$$

The regional local fixed demands for grain are presented in the last column of Table A3.12. These values are the RHS values of the first 26 constraints in all the models of Chapter A2.

Appendix 3

Table A3.12   Estimations of local fixed demands for grain, 1985

j	Livestock prod.('000 ton)			Draft animal ('000 head)	Grain requirements ('000 ton)					
	Milk (1)	Egg (2)	Meat (3)	(4)	Seed (5)	Milk (6)	Egg (7)	Meat (8)	Anim. (9)	Total (10)
1	281	581	1085	4965	837	196	1626	3255	1042	6958.80
2	102	110	224	2604	296	71	308	672	546	1894.41
3	259	91	359	7366	217	181	254	1077	884[a]	2614.50
4	102	304	614	3033	351	71	851	1842	636	3752.89
5	64	172	320	2946	441	44	481	960	618	2546.17
6	454	205	349	3055	478	317	574	1047	641	3059.15
7	142	104	197	68	77	99	291	591	14	1072.85
8	75	607	1603	853	583	52	1699	4809	179	7323.90
9	107	162	836	761	1125	74	453	2508	159	4321.85
10	19	247	866	4509	780	13	691	2598	946	5030.27
11	42	83	497	1186	286	29	232	1491	249	2287.84
12	18	114	643	2485	552	12	319	1929	521	3334.71
13	152	725	1286	4423	1129	106	2030	3858	928	8052.80
14	45	371	718	8864	975	31	1038	2154	1861	6061.52
15	38	401	1016	3257	797	26	1122	3048	684	5679.17
16	10	244	1478	3505	905	7	683	4434	736	6765.40
17	43	126	1395	5210	625	30	352	4185	1094	6287.64
18	6	41	723	5827	402	4	114	2169	1223	3913.83
19	223	325	3012	9868	1379	156	910	9036	2072	13553.43
20	6	46	505	5460	214	4	128	1515	1146	3008.80
21	147	64	665	14135	355	102	179	1995	2968	5601.17
22	146	112	302	2556	342	102	313	906	536	2201.24
23	44	49	248	5079	191	30	137	744	1066	2169.57
24	159	9	112	5894	36	111	25	336	472[b]	980.12
25	13	11	34	733	50	9	30	102	153	346.05
26	197	43	178	5176	179	137	120	534	621[c]	1592.98
Sum	2894	5347	19265	113818	—	—	—	—	—	110447.07

Notes:
(5) = 0.036×(total outputs of grain) (Table A3.9)
(6) = 0.7×(1), (7) = 2.8×(2), (8) = 3.0×(3),
(9) = 0.21×(4).
[a] for region 3 (In.Mon.) (9) = 0.12×(4).
[b] for region 24 (Qinghai) (9) = 0.08×(4).
[c] for region 26 (Xinjiang) (9) = 0.12×(4).
Source: The first four columns, SSB, CAYB (1986) Livestock.

A3.3 REGIONAL SUPPLY FUNCTIONS

A3.3.1 Production costs

It is assumed that the production costs for all regions are exactly equal to the estimated demand prices presented in the last section. Therefore, for the basic model described in problem A2.1 and the quota model in problem A2.2, the supply or production costs are the same as APP_1 in Table A3.10.

The production costs for the consumer subsidy models described in problems A2.3 and A2.4 are assumed to be in between the mixed average procurement prices APP and the estimated demand prices APP_1 (Table A3.10), and they are assumed to be 25 per cent higher than the average procurement prices APP. Therefore APP_2 in Table A3.10 are used in problems A2.3 and A2.4. The choice of APP_2 is arbitrary, but it is based on the real demand and supply structure in China.

With consumer subsidies, about 20 per cent of the total population (the urban residents) pays a price for grain much lower than the mixed APP, while 80 per cent of the population (the rural residents) has to pay the free market price. Thus, the average consumer price must be in between these two extremes. The choice of APP_2 is an approximation for such an average price. To understand the effects of consumer subsidies, two points must be made clear: (i) The closer APP_2 is to APP_1, the less effects of consumer subsidies; (ii) The lower the values of APP_2, the more subsidies urban consumers will receive and the greater effects of subsidies.

For the production tax models in problems A2.5 and A2.6, production costs are assumed to be APP_3 in Table A3.10, which are 70 per cent higher than the mixed APP. The reason for that is explained as follows.

If real production costs are APP_1 (40 per cent higher than APP) in Table A3.10, with a production tax, the total costs of production will be higher. Therefore, the choice of APP_3, which is higher than APP_1, is arbitrary, but a realistic approximation of the 'true' production costs with a production tax. It is obvious that the higher the values of APP_3, the greater the effects of a production tax, and vice versa.

A3.3.2 Land productivity and constraints

The models presented in Chapter A2 assume that land productivity of grain is the average productivity of those in 1984 and 1985 for all

Table A3.13   Land productivity and constraints of grain crops, 1985

j	Land productivity (ton/mu)				Arable land ('000mu)		
	1984 (1)	*1985* (2)	*Average* (3)	*1/(3)* (4)	*1985* (5)	*UB* (6)	*LB* (7)
1	0.1932	0.2082	0.2007	4.9824	1117.51	1162.21	1005.76
2	0.1745	0.1800	0.1772	5.6418	458.26	476.59	412.43
3	0.1055	0.1180	0.1117	8.9486	513.27	533.80	461.94
4	0.3065	0.2250	0.2658	3.7629	433.43	450.77	390.09
5	0.3110	0.2490	0.2800	3.5714	492.52	512.22	443.27
6	0.1595	0.1320	0.1457	6.8611	1082.46	1125.76	974.21
7	0.3460	0.3260	0.3360	2.9762	65.74	68.37	59.17
8	0.3405	0.3240	0.3322	3.0098	964.87	1003.46	868.38
9	0.3480	0.3300	0.3390	2.9499	490.68	510.31	441.61
10	0.2370	0.2450	0.2410	4.1494	884.79	920.18	796.31
11	0.2810	0.2810	0.2810	3.5587	283.27	294.60	254.94
12	0.2780	0.2800	0.2790	3.5842	547.64	569.55	492.88
13	0.2585	0.2620	0.2602	3.8425	1197.64	1245.55	1077.88
14	0.2145	0.2000	0.2073	4.8251	1354.40	1408.58	1218.96
15	0.2850	0.2890	0.2870	3.4843	766.24	796.89	689.62
16	0.3230	0.3250	0.3240	3.0864	774.21	805.18	696.79
17	0.2715	0.2590	0.2653	3.7700	669.82	696.61	602.84
18	0.2190	0.2160	0.2175	4.5977	517.09	537.77	465.38
19	0.2775	0.2721	0.2748	3.6389	1408.35	1464.68	1267.51
20	0.2180	0.1790	0.1985	5.0378	331.81	345.08	298.63
21	0.1930	0.1875	0.1903	5.2558	526.89	547.97	474.20
22	0.1700	0.1597	0.1649	6.0653	594.84	618.63	535.36
23	0.1275	0.1270	0.1273	7.8585	416.23	432.88	374.61
24	0.1655	0.1730	0.1693	5.9084	57.99	60.31	52.19
25	0.1510	0.1430	0.1470	6.8027	97.57	101.47	87.81
26	0.1665	0.1790	0.1727	5.7887	279.25	290.42	251.32

Notes:

$(3) = \frac{1}{2}[(1) + (2)]$; $(4) = 1/(3)$; $(6) = 1.04 \times (5)$

(5) is the actual areas sown to grain crops in 1985;

$(7) = 0.90 \times (5)$.

Sources: Column 1, SSB, CAYB (1985), p. 265. Columns 2 and 5 SSB, CAYB (1986) Agriculture. Other columns are derived from columns 1, 2 and 5.

regions. For each region, the productivity in 1984 and 1985 and their averages are presented in the first three columns of Table A3.13. The values in the fourth column are the inverse of the average land productivity in column 3. Column 5 presents the actual areas sown to grain crops in 1985 for all regions. The values in column 6 are 0.04 per cent higher than those in column 5. They are the upper bound constraints of land. The lower bound constraints of land are 90 per cent of those in column 5.

A4 Results and Policy Implications

This chapter reports the simulation results from the models of Chapter A2 using the data in Chapter A3. The welfare impacts upon producers, consumers and the budget of different policy scenarios embodied in the models are analysed with some necessary assumptions. Special analysis is focused on the impact of the policy scenarios on the patterns of grain production.

The models are solved with a computer package called LINDO, a specialized package for linear and non-linear optimization problems. As the objective functions of all the models are quadratic, the solution matrix must be in a non-linear form. To form the solution matrix, a conventional Lagrangian equation is established to derive the first-order Kuhn–Tucker conditions. As all constraints are linear, the first-order conditions simultaneously satisfy the sufficient conditions. Thus, the first-order conditions are able to produce a unique solution to all the variables, multipliers and shadow prices.

According to the neoclassical economic theory, government intervention via whatever policy instruments will have two consequences: (1) causing welfare redistribution among different interest groups; and (2) incurring an efficiency loss to the society. In this chapter, welfare redistribution and efficiency loss are explained indirectly. That is, government intervention will distort the pattern of specialized production, hence resulting in an efficiency loss in production.

This chapter is divided into three sections. Section one reports the simulation results of the models, including the results of regional supplies, demands, inter-regional flows, prices and the shadow prices of land constraints. The second section analyses the welfare effects of different policy scenarios, and their impact on the patterns of grain production. The last section discusses policy implications of government intervention.

To make it more clear, the models in Chapter A2 are briefly restated here.

Model A2.1: the 'basic' model under the assumption of perfect competition without any government intervention.

272

Model A2.2: a compulsory production model under the assumption that each region is required to grow a minimum acreage of grain (90 per cent of the actual crop area in 1985).

Model A2.3: a consumer subsidy model in which regional production costs are artificially brought down by a unit product subsidy so that grain prices are brought down to meet the objective of consumer subsidy.

Model A2.4: a consumer subsidy and compulsory production quota model.

Model A2.5: a production tax model in which a unit product tax is levied.

Model A2.6: a production tax and compulsory production quota model in which producers in each region have to pay a unit product tax and meet the minimum acreage quota.

A4.1 COMPUTATIONAL RESULTS

A4.1.1 Notations

In this chapter, the following notation will be used in order to simplify the analysis.

$S1_i$, $S2_i$, ... $S6_i$ are respectively the optimal equilibrium supplies in region i obtained from models A2.1 to A2.6.

$D1_i$, $D2_i$, ... $D6_i$ are respectively the optimal equilibrium demands in region i obtained from models A2.1 to A2.6.

$P1_i$, $P2_i$, ... $P6_i$ are respectively the optimal equilibrium market prices in region i obtained from models A2.1 to A2.6.

$Q1_i$, $Q2_i$, ... $Q6_i$ are respectively the optimal equilibrium shadow prices of land with respect to the land upper bound constraints in region i obtained from models A2.1 to A2.6.

$E2_i$, $E4_i$, $E6_i$ are respectively the optimal equilibrium shadow prices of land with respect to the land lower bound constraints in region i obtained from models A2.2, A2.4 and A2.6.

$D0_i$ is the fixed local demand for grain in region i (Chapter A3).

A4.1.2 The results

The simulation results from models A2.1 to A2.6 are reported in Tables A4.1 to A4.10. The first 4 tables report the optimum equilibrium supplies, demands, prices and shadow prices of land lower and upper bound constraints by region. The first columns of these tables present the regional numbers ($j = 1, 2, \ldots, 26$). Tables A4.5 to A4.10 report the inter-regional flows of grain obtained from the models. In each table, the regional number labelled with i in the first row represents the outflow (supply) regions. The regional number labelled with j in the first column represents the inflow (demand) regions. The total amounts of inter-regional flows and their shares as a percentage of total supplies for all models are reported in Table A4.1.

The results of model A2.1 are used as the baseline for comparison as it assumes perfect competition without any government intervention. Following the norm of neoclassical economics, Model A2.1 is regarded as the most efficient model. Any violation of this model will induce some distortion and efficiency loss. The analysis in the next section will, therefore, compare the results of all the other models with those of model A2.1.

A4.2 POLICY ANALYSIS

This section is divided into six sub-sections. The first sub-section will discuss the simulation results of regional optimal grain supplies and prices from model A2.1. The last five sub-sections will describe the impacts of each policy scenario on welfare distribution and the patterns of grain production.

A4.2.1 Simulation results of model A2.1

According to the values of Q1 in Table A4.4, supply regions in model A2.1 can be divided into two groups, with one consisting of regions with positive shadow prices and the other consisting of regions with zero shadow prices of land upper bound constraints. Thus, optimal regional grain supplies and equilibrium regional prices may be described as follows:

Table A4.1 Optimal supplies of grain ('000 tons)

J	S1	S2	S3	S4	S5	S6
Beijing	15220.54	20157.94	18029.33	20157.94	8887.57	20157.94
Shanxi	8445.54	7892.54	8445.54	8445.54	8445.54	7308.57
In.Mong.	5946.57	5946.57	5946.57	5946.57	5946.57	5946.57
Liaoning	11697.68	11697.68	11697.68	11697.68	11697.68	10123.00
Jilin	14166.16	14166.16	14166.16	14166.16	14166.16	13193.02
Heilongjiang	16261.86	14072.69	16261.86	16261.86	16261.86	14072.69
Shanghai	2295.23	1986.04	2295.22	2295.22	2295.22	1986.04
Jiangsu	25931.09	28834.51	28163.40	28834.51	22952.26	28834.51
Zhejiang	17287.51	15112.20	17287.51	17287.51	16729.13	14960.20
Anhui	18964.51	19185.88	19738.96	19738.96	17415.60	19185.88
Fujian	8278.30	8278.30	8278.30	8278.30	8278.30	7163.85
Jiangxi	15890.13	14077.43	15890.13	15890.13	12993.33	13750.80
Shangdong	25188.70	28050.80	26346.92	28050.80	22872.27	28050.80
Henan	21981.19	25232.04	27661.71	25232.04	11608.05	25232.04
Hubei	22869.57	22274.34	22869.57	22869.57	22869.57	19790.79
Hunan	26088.00	26088.00	26088.00	26088.00	26088.00	26088.00
Guangdong	18467.43	18467.43	18467.43	18467.43	18467.43	18467.43
Guangxi	11695.99	11695.99	11695.99	11695.99	11695.99	11695.99
Sichuan	40246.20	40246.20	40246.20	40246.20	40246.20	38889.55
Guizhou	6783.70	6783.70	6783.70	6783.70	6783.70	6783.70
Yunnan	10423.82	10423.82	10423.82	10423.82	10423.82	10423.82
Shaanxi	10189.58	10189.58	10189.58	10189.58	10189.58	8817.86
Gansu	5508.36	5508.36	5508.36	5508.36	5508.36	4766.88
Qinghai	1020.25	1020.25	1020.25	1020.25	1020.25	882.89
Ningxia	1490.50	1289.84	1490.50	1490.50	1317.80	1289.84
Xinjiang	4563.59	4563.59	4785.34	4785.34	4119.03	4335.87
$\sum$	366901.99	373241.89	379778.02	381851.95	339279.27	362198.53
TR	25451.40	21673.13	24540.59	23540.85	30980.70	23415.85
RS(%)	6.94	5.81	6.46	6.16	13.54	9.30

Notes:
$\sum$ = Total grain supply of all regions. TR = Total amount of inter-regional flows. RS = Share of total inter-regional flow in total supply. J = regional numbers. S1 to S6 are optimal regional supply vectors for models A2.1 to A2.6.

Table A4.2   Optimal demands for grain ('000 tons)

J	D1	D2	D3	D4	D5	D6
Beijing	16284.37	17017.96	17388.61	17664.56	14077.80	16209.08
Shanxi	5801.45	5998.13	6164.33	6234.24	5114.47	5683.03
In.Mong.	4695.03	4896.08	4997.67	5073.30	4128.89	4674.39
Liaoning	8947.91	9317.20	9503.79	9642.70	7875.02	8910.01
Jilin	6031.56	6261.72	6378.01	6464.59	5339.25	6007.94
Heilongjiang	8402.45	8734.32	8902.01	9026.85	7404.19	8368.38
Shanghai	3463.07	3554.83	3644.98	3666.65	3114.66	3424.23
Jiangsu	17487.38	17956.23	18416.89	18527.60	15628.36	17288.90
Zhejiang	11669.08	11703.08	12082.09	12127.67	10514.98	11270.49
Anhui	13934.24	14157.06	14708.70	14708.70	12385.33	13764.31
Fujian	6852.15	6909.37	7130.15	7160.83	6108.94	6617.48
Jiangxi	10488.14	10561.13	10842.75	10881.88	9540.15	10188.80
Shangdong	17135.90	17967.55	18294.12	18646.94	14819.47	17139.82
Henan	18372.77	19092.28	19567.89	19695.81	15984.55	18262.87
Hubei	13878.02	14065.28	14396.42	14452.19	12350.06	13559.43
Hunan	16535.55	16734.07	17111.33	17174.87	15144.44	16157.78
Guangdong	14600.50	14821.25	15240.78	15311.42	13053.54	14180.40
Guangxi	8371.60	8508.48	8768.59	8812.40	7774.06	8111.14
Sichuan	23682.80	24043.01	24386.16	24501.44	21693.55	22997.31
Guizhou	5634.35	5739.20	5938.46	5972.02	5055.32	5434.82
Yunnan	7131.37	7258.75	7380.09	7420.85	6427.95	6888.97
Shaanxi	7326.11	7555.54	7740.67	7820.52	6629.91	7206.59
Gansu	4796.50	4952.58	5078.53	5132.85	4377.77	4732.96
Qinghai	911.32	942.47	967.60	978.45	827.76	898.64
Ningxia	1086.73	1112.64	1144.01	1149.25	971.75	1066.84
Xinjiang	2970.61	2970.61	3192.36	3192.36	2526.05	2742.89
$\sum$	256490.96	262830.82	269366.98	271440.92	228868.20	251787.49

Notes:
$\sum$ = Total national demand. The amounts of regional optimum demand do not include regional fixed demand (Chapter A3). If total fixed demand is included, total demand in this table must be equal to total corresponding supply in Table A4.1. J denotes regional numbers. $D1$ to $D6$ are optimal regional demand vectors for models A2.1 to A2.6.

Table A4.3 Optimal equilibrium prices (¢/kg)

J	P1	P2	P3	P4	P5	P6
Beijing	53.37	49.57	47.65	46.22	64.80	53.76
Shanxi	53.78	50.94	48.54	47.53	63.70	55.49
In.Mong.	55.22	51.42	49.50	48.07	65.92	55.61
Liaoning	51.02	47.22	45.30	43.87	62.06	51.41
Jilin	49.59	45.79	43.87	42.44	61.02	49.98
Heilongjiang	48.77	44.97	43.05	41.62	60.20	49.16
Shanghai	53.95	51.09	48.28	47.60	64.81	55.16
Jiangsu	52.95	50.09	47.28	46.60	64.29	54.16
Zhejiang	52.47	52.15	48.58	48.15	63.33	56.22
Anhui	53.20	51.56	47.50	47.50	64.60	54.45
Fujian	54.61	53.81	50.72	50.29	65.00	57.89
Jiangxi	52.94	52.14	49.05	48.62	63.33	56.22
Shangdong	53.27	49.17	47.56	45.82	64.69	53.25
Henan	54.87	51.33	48.99	48.36	66.62	55.41
Hubei	52.81	51.37	48.82	48.39	64.56	55.26
Hunan	53.94	52.60	50.05	49.62	63.33	56.49
Guangdong	56.49	55.15	52.60	52.17	65.88	59.04
Guangxi	57.28	55.94	53.39	52.96	63.13	59.83
Sichuan	54.84	53.50	52.22	51.79	62.24	57.39
Guizhou	57.39	56.05	53.50	53.07	64.79	59.94
Yunnan	57.92	56.58	55.30	54.87	65.32	60.47
Shaanxi	55.86	52.96	50.62	49.61	64.66	57.37
Gansu	58.14	55.24	52.90	51.89	65.92	59.32
Qinghai	59.00	56.10	53.76	52.75	66.78	60.18
Ningxia	56.29	53.92	51.05	50.57	66.81	58.11
Xinjiang	57.74	57.74	51.56	51.56	70.13	64.09

Notes:
J denotes regional numbers. P1 to P6 are optimal regional price vectors for models A2.1 to A2.6.

Appendix 4

Table A4.4   Shadow prices of land constraints

J	Q1	Q2	E2	Q3	Q4	E4	Q5	Q6	E6
Beijing	0.000	0.000	0.763	0.000	0.000	0.287	0.000	0.000	2.216
Shanxi	0.503	0.000	0.000	0.542	0.363	0.000	0.328	0.000	1.127
In.Mong.	1.193	0.769	0.000	1.088	0.929	0.000	1.323	0.171	0.000
Liaoning	1.860	0.850	0.000	1.595	1.215	0.000	2.288	0.000	0.542
Jilin	2.360	1.296	0.000	1.994	1.593	0.000	3.091	0.000	0.000
Heilongjiang	0.334	0.000	0.220	0.226	0.018	0.000	0.548	0.000	1.061
Shanghai	0.581	0.000	0.380	0.554	0.327	0.000	0.470	0.000	2.772
Jiangsu	0.000	0.000	0.950	0.000	0.000	0.224	0.000	0.000	3.365
Zhejiang	0.108	0.000	0.000	0.686	0.541	0.000	0.000	0.000	2.410
Anhui	0.000	0.000	0.395	0.000	0.000	0.000	0.000	0.000	2.446
Fujian	0.719	0.495	0.000	1.192	1.072	0.000	0.503	0.000	1.495
Jiangxi	0.223	0.000	0.000	0.696	0.576	0.000	0.000	0.000	1.984
Shangdong	0.000	0.000	1.067	0.000	0.000	0.453	0.000	0.000	2.977
Henan	0.000	0.000	0.734	0.000	0.000	0.130	0.000	0.000	2.323
Hubei	0.413	0.000	0.000	0.851	0.727	0.000	0.629	0.000	2.041
Hunan	2.589	2.155	0.000	2.924	2.785	0.000	2.443	0.227	0.000
Guangdong	3.013	2.658	0.000	3.266	3.152	0.000	2.942	1.127	0.000
Guangxi	2.997	2.706	0.000	3.165	3.072	0.000	2.242	1.525	0.000
Sichuan	2.083	1.715	0.000	2.755	2.637	0.000	1.333	0.000	0.000
Guizhou	2.612	2.346	0.000	2.780	2.695	0.000	2.197	1.235	0.000
Yunnan	2.217	1.962	0.000	2.662	2.581	0.000	1.737	0.814	0.000
Shaanxi	0.631	0.153	0.000	0.688	0.521	0.000	0.246	0.000	0.956
Gansu	0.506	0.137	0.000	0.578	0.449	0.000	0.019	0.000	0.821
Qinghai	0.706	0.215	0.000	0.814	0.643	0.000	0.036	0.000	1.081
Ningxia	0.187	0.000	0.162	0.282	0.212	0.000	0.000	0.000	1.279
Xinjiang	0.000	0.000	0.000	0.000	0.000	0.000	0.000	0.000	1.044

Notes:
Q1 to Q6 are respectively the shadow price vectors with respect to land upper bound constraints in models A2.1 to A2.6. J denotes regional numbers. E2, E4 and E6 are respectively the shadow price vectors with respect to land lower bound constraints in models A2.2, A2.4 and A2.6.

Table A4.5 Interregional flows (x_{ij})(100t) from model A2.1

	Samxi	Jiln	Heil	Jisu	Zeja	Jaxi	Hube	Huma	Sicn	Saam	Nixa
Beijing		32224	48003								
In.Mong.		13630									
Liaoning		10031									
Shanghai				11198	11209						
Fujian					1757	6860					
Henan							24531				
Guangdong						13813		10394			
Guangxi								5894			
Guizhou								11582	7013		
Yunnan									23087		
Gansu	7497						7378			662	
Qinghi							1215				577

Table A4.6 Inter-regional flows (x_{ij})(100t) from model A2.2

	Jiln	Heil	Jisu	Jaxi	Sadn	Hena	Hube	Huma	Sicn	Saam
Beijing	15396	22792								
In.Mong.	15640									
Liaoning	13724		26416							
Shanghi			9127							
Zhejiang					15					
Anhui					7373					
Fujian				1816						
Guangdong							25299	1116		
Guangxi								7263		
Guizhou								17506	2137	
Yunnan						782			24361	
Gansu	7134				3893					
Qinghai					9023					4328
Ningxia	1688									

Notes: i and j respectively denote the outflow and inflow regional numbers.

Table A4.7 Inter-regional flows (x_{ij})(100t) from model A2.3

	Sanx	Jiln	Heil	Jisu	Zeja	Jixi	Hena	Hube	Huna	Sicn	Saan	Nixa
Beijing		20174	43007									
In.Mong.		16656										
Liaoning		15590		24226								
Shanghai					8836							
Fujian						2561						
Guangdong						2670		27940				
Guangxi									9864			
Guizhou						9387			12248			
Yunnan						2508				23066		
Gansu	3868						14916					
Qinghai							5407				2477	4

Table A4.8 Inter-regional flows (x_{ij})(100t) from model A2.4

	Sanx	Jiln	Heil	Jisu	Zeja	Jixi	Sndn	Hube	Huna	Sicn	Saanx
Beijing		2896	41759								
In.Mong.		17412									
Liaoning		16979		24443							
Shanghai					8380						
Fujian						3324					
Henan		5253		5387							
Guangdong						9343		16586			
Guangxi									10302		
Guizhou								10796	11175		
Yunnan						4069				21913	
Gansu	3169	8966					4127				
Qinghai							9383				1678
Ningxia		48									

Notes: *i* and *j* denote respectively the outflow and inflow regional numbers.

Table A4.9 Inter-regional flows (x_{ij})(100t) from model A2.5

	Sanx	Liao	Jiln	Heil	Zeja	Jixi	Hube	Huna	Gnxi	Sicn	Saanx
Beijing	7968										
In.Mong.		698	62807	57985							
Shanghai					18923						
Fujian						1185					
Henan	6398						48403	33125		2869	13584
Guangdong								8656	81		
Guizhou										12804	
Yunnan										16053	
Gansu										10390	
Qinghai										7876	

Table A4.10 Inter-regional flows (x_{ij})(100t) from model A2.6

	Jiln	Heil	Sahi	Jisu	Anhu	Jixi	Sgdn	Hena	Hube	Huna	Sich
Beijing	3648										
Shanxi	2689										
In.Mong.	13423	26452									
Liaoning	25399										
Shanghai				39809							
Zhejiang				2408	3913						
Fujian						2273					
Guangdong			14698				444				20006
Guangxi										3290	2725
Guizho									5522	8352	
Yunnan										20663	
Shaanxi											
Gansu							21356	5900			
Qinghai							6782	3177			
Ningxia	1230										

Notes: i and j denote respectively the outflow and inflow regional numbers.

(1) Group one $(Q1_i > 0)$

(i) *Regional grain supplies*
This group has twenty regions, including Shanxi, Inner Mongolia, Liaoning, Jilin, etc. (Table A4.4). Total grain supplies of these regions are all restricted by land upper bound constraints. Thus,

$$r_i S1_i = L_i \qquad\qquad\qquad\qquad \text{(A4.1)}$$

where r_i, $S1_i$ and L_i are respectively the inverse of land productivity of grain, the optimal grain supply and the land upper bound constraint in region i.

The value of $Q1_i$ represents the opportunity cost of introducing an extra unit of land into grain production in region i. The greater the value of $Q1_i$, the greater the opportunity cost of an extra unit of land in region i.

As discussed in Chapter A2, regions with higher land shadow prices with respect to the upper bound constraints are usually those with lower production costs, but this does not mean that these regions have a greater comparative advantage in grain production because they may have the same advantage in the production of non-grain crops as well (detailed discussion, see Chapter A2).

(ii) *Regional grain market prices*
Optimal equilibrium grain prices in these regions are greater than their production costs. The relationship between the optimal market price and production cost in a particular region *is* as follows:

$$P1_i = P_{si} + r_i Q1_i \qquad\qquad\qquad \text{(A4.2)}$$

where $P1_i$ and $Q1_i$ are respectively the optimal market price and the shadow price of land in region i. P_{si} and r_i are respectively the production cost and inverse of land productivity in region i.

(2) Group two $(Q1_i=0)$

(i) *Regional grain supplies*
This group consists of six regions, including Beijing, Jiangsu, Anhui, Shangdong, Henan and Xinjiang (Table A4.4). Because the shadow prices of land with respect to upper bound constraints are zero, total optimal grain supplies in these regions are not restricted by arable land. Thus,

$$r_i S1_i \geq L_i \tag{A4.3}$$

where r_i, $S1_i$ and L_i are defined above.

This implies that the optimal areas sold to grain crops in these regions may be less than the maximum land available for grain production.

(ii) *Regional grain market prices*

Equilibrium grain market prices in these regions are exactly the same as their production costs. Thus,

$$P1_i = P_{si} \tag{A4.4}$$

where $P1_i$ and P_{si} are defined above.

A4.2.2 Effects of production quotas (model A2.2)

The simulation results of model A2.2 are presented in Table A4.4. These results are compared with the results of the basic model (model A2.1).

(1) Welfare effects

From Tables A4.2 and A4.4, the regional optimal demands for grain from model A2.2 are greater than or at least equal to those from model A2.1 for all regions while the optimum regional prices in model A2.2 are lower than or equal to those in model A2.1. Therefore, with a production quota consumers in all regions will benefit. Let $CGN2_i$ represent the total consumer gain in region i due to a production quota, then

$$CGN2_i = \tfrac{1}{2}(D1_i + D2_i + 2D0_i)(P1_i - P2_i) \quad \text{for all } i \tag{A4.5}$$

From Tables A4.3 and A4.4, the producers in all regions except region 26 (Xinjiang) will lose as a result of a production quota. The producers' loss varies significantly in different regions, depending on the shadow prices of the lower and upper bound land constraints. To calculate producer losses, supply regions are classified into four groups according to the shadow prices $Q1_i$, $Q2_i$ and $E2_i$.

(i)   Group 1: $Q1_i = Q2_i = 0, E2_i > 0$. This implies that total supplies of these regions in both models A2.1 and A2.2 do not reach the upper bound constraints of land, i.e. $r_i S1_i < L_i$ and $r_i S2_i < L_i$, but they are restricted by the lower bound constraints, i.e. $r_i S2_i = I_i$. In other words, they are forced to produce at least the minimum quota.

(ii)   Group 2: $Q1_i > 0$, $Q2_i > 0$, $E2_i = 0$. This implies that total supplies of these regions in both models reach their upper bound constraints of land, i.e., $r_i S1_i = r_i S2_i = L_i$. In terms of production, a compulsory quota does not affect these regions at all.

(iii)   Group 3: $Q1_i > 0$, $E2_i > 0$, $Q2_i = 0$. This implies that total supplies of these regions reach their upper bound constraints of land in model A2.1 but not in model A2.2. However, they are restricted by the lower bound constraints of land in model A2.2. Mathematically, $r_i S1_i = L_i$ and $r_i S2_i = I_i$.

(iv)   Group 4: $Q1_i > 0$, $E2_i = Q2_i = 0$. This implies that the supplies of these regions are restricted by the upper bound constraints of land in model A2.1 but not restricted by either the lower or upper bound constraints of land in model A2.2, i.e., $I_i \geq r_i S2_i \geq L_i$ and $r_i S1_i = L_i$.

Let $PLS2_i$ represent the loss to the producers in region i from model A2.2 as opposed to model A2.1. Then,

$$PLS2_i = (P1_i - P2_i)S2_i + r_i Q1_i(S1_i - S2_i) \qquad \text{for all } i \qquad (A4.6)$$

As for the net loss to each individual region, it is the difference between the gain to the consumers and the loss to the producers in the same region. Let the net loss to region i be denoted by $NLS2_i$ (positive gain, negative loss, the following will be the same), then

$$NLS2_i = CGN2_i - PLS2_i \qquad \text{for all } i \qquad (A4.7)$$

Regional losses and gains are presented in columns 2 to 4 in Table A4.11. The total producers' loss is about 8.7 billion yuan while the total consumers' gain is about 8.4 billion yuan. The net loss to the society (efficiency loss) is about 273 million yuan.

The efficiency loss is not tòo significant, but the producers' loss and consumers' gain are substantial. If the variations of loss in different regions are ignored, a compulsory production quota is an effective policy method to extract welfare from the producers to favour consumers without incurring a substantial efficiency loss at the national level. If regional factors are taken into account, the situation becomes more complicated. As seen from column 3 in Table A4.11, some regions, such as Jilin, Heilongjiang, Jiangsu and Shangdong suffer large losses, whereas other regions, such as Beijing, Inner Mongolia, Liaoning, Shanghai and Guangdong, have significant net gains.

Table A4.11   Welfare analysis for model A2.2 (production quota)

j	PSL2 (1000¥)	CGN2 (1000¥)	NLS2 (1000¥)	S21 ('00t)	D21 ('00t)	P21 (¢/kg)
Beijing	766000	897177	131176	49374	7335	−3.80
Shanxi	239841	221355	−18486	−5530	1966	−2.84
In.Mong.	225969	281581	55612	0	2010	−3.80
Liaoning	444512	489647	45135	0	3692	−3.80
Jilin	538314	330326	−207987	0	2301	−3.80
Heilongjiang	584929	441846	−143083	−21891	3318	−3.80
Shanghai	62147	131039	68892	−3091	917	−2.86
Jiangsu	824667	716307	−108359	29034	4688	−2.86
Zhejiang	55289	51225	−4064	−21753	340	−0.32
Anhui	314648	312844	−1803	2213	2228	−1.64
Fujian	66226	73348	7122	0	572	−0.80
Jiangxi	127107	110874	−16233	−18127	729	−0.80
Shangdong	1150082	1049784	−100297	28621	8316	−4.10
Henan	893214	877709	−15505	32508	7195	−3.54
Hubei	329316	282972	−46344	−5952	1872	−1.44
Hunan	349578	313561	−36016	0	1985	−1.34
Guangdong	247464	281380	33916	0	2207	−1.34
Guangxi	156726	165542	8815	0	1368	−1.34
Sichuan	539300	501380	−37920	0	3602	−1.34
Guizhou	90901	116521	25619	0	1048	−1.34
Shaanxi	139679	171469	31790	0	1273	−1.34
Shaanxi	295497	279619	−15877	0	2294	−2.90
Gansu	159742	204279	44536	0	1560	−2.90
Qinghai	29587	55303	25716	0	311	−2.90
Ningxia	33121	34263	1142	−2006	259	−2.37
Xinjiang	0	0	0	0	0	0.00
Σ	8663865	8391362	−272503	63398	63398	—

Notes:
$PLS2_i = (P1_i − P2_i)S2_i + r_iQ1_i(S1_i − S2_i),$
$CGN2_i = \frac{1}{2}(D2_i + D1_i + 2D0_i)(P1_i − P2_i),$
$NLS2_i = CGN2_i − PLS2_i,$
$S21_i = S2_i − S1_i,$
$D21_i = D2_i − D1_i,$
$P21_i = P2_i − P1_i,$

$PLS2_i$ is the producer loss in region i: A positive sign means loss; a negative sign means gain.
$GN2_i$ is the consumer gain in region i: A positive sign means gain; a negative sign means loss.
$NLS2$ is the net loss in region i: A negative sign means loss; a positive sign means gain.
$S21$, $D21$ and $P21$ are respectively the regional differential vectors of supply, demand and price between model A2.2 and model A2.1.

(2) Effects on the production pattern

Regional differentials of equilibrium supplies, demands, and prices between models A2.1 and A2.2 are presented in the last three columns of Table A4.11. With quotas, total national production increases by about 6.3 million tons, or 1.73 percent. The change in production at the regional level varies significantly. Some regions may even produce less or maintain the same output level. Regions with increased production are those with relatively high production costs, whilst regions with reduced production are those with relatively low production costs. Consequently, apart from increasing total production, a production quota distorts the spatial pattern of grain production in such a way that the high-cost regions increase their supplies while the low-cost regions reduce theirs, resulting in an efficiency loss to the society.

A4.2.3 Effects of consumer subsidies (model A2.3)

Consumer subsidy is defined as supplying consumers (urban residents) with foodgrains at a low price. In model A2.3, because regional supply functions are horizontal, it is assumed that grain producers receive a product subsidy so that the cost of production is reduced from APP1 to APP2 (Chapter A3). A reduction in production costs will, therefore, reduce the market equilibrium prices. Consequently, if all the producers and consumers receive and pay the new equilibrium prices, the objective of consumer subsidy will be materialized. That is, consumers benefit from paying a price lower than the free market price – the equilibrium price of the basic model (model A2.1).

(1) Welfare effects

From Tables A4.2 and A4.3, in comparison with the results of model A2.1, regional demands from model A2.3 are greater and regional prices are lower. Therefore, the consumer surplus in every region *i*ncreases. The consumers' gain in region i (denoted by $CGN3_i$), is

$$CGN3_i = \tfrac{1}{2}(D1_i + D3_i + 2D0_i)(P1_i - P3_i) \qquad \text{for all } i \qquad (A4.8)$$

Compared with the results from model A2.1, the results from model A2.3 show that producers gain in some regions but lose or breakeven in others under this policy scenario. Why producers in some regions lose may be explained as follows: The introduction of a subsidy on production will increase total grain supply but depress prices. Thus, producers in some regions (usually the low-cost and surplus regions) may have more loss due to price depression than gain from the subsidy.

From Table A4.4, supply regions can be classified into two groups in order to calculate the changes in producers' surplus by region.

(i) Group 1: $Q1_i > 0$ and $Q3_i > 0$. This implies that total supplies in both models A2.1 and A2.3 reach the upper bound constraints of land, i.e. $r_i S1_i = r_i S3_i = L_i$.

(ii) Group 2: $Q1_i = Q3_i = 0$, or $P1_i = P_{si}$ and $P3_i = P_{ci}$. This implies that $r_i S1_i \leq L_i$, and $r_i S3_i \leq L_i$.

Where P_{si} and P_{ci} are respectively the production costs in region i for models A2.1 and A2.3. And $P_{si} = \text{APP1}_i$, $P_{ci} = \text{APP2}_i$ (Chapter A3).

Let $PGN3_i$ represent the producers' gain in region i in model A2.3 as opposed to the results in model A2.1. Then, the producers' gains can be calculated according to the following formulae:

Group 1: $PGN3_i = r_i(Q3_i - Q1_i)S3_i$ (A4.9)

Group 2: $PGN3_i = r_i(Q3_i - Q1_i)S3_i = 0$ (A4.9a)

The budgetary cost for each region can be denoted by $BCS3_i$, and

$$BCS3_i = (P_{si} - P_{ci})S3_i \qquad \text{for all } i \qquad\qquad\text{(A4.10)}$$

Let the net loss to region i be denoted by $NLS3_i$, then

$$NLS3_i = PGN3_i + CGN3_i - BCS3_i \qquad \text{for all } i \qquad\text{(A4.11)}$$

The gains (losses) to producers, consumers, the budget and the net losses by region are presented in columns 2 to 5 of Table A4.12. Producers in In.Mon, Liaoning, Jilin, Heilongjiang and Shanghai lose but gain or breakeven in the rest of country. Consumers in all regions have substantial gains. The total gain to consumers is about 17.4 billion yuan; the total gain to producers is only 2.5 billion yuan; but the total budgetary cost is more than 20.2 billion yuan. The total efficiency loss to the society is 330 million yuan.

(2) Impact on the production pattern

The supply and demand differentials, $S31_i$ and $D31_i$, and the optimum price differentials $P31_i$ are presented in the last three columns of Table A4.12. The total supply and demand increase by about 13 million tons, or 3.51 per cent. The increase of supply is concentrated in a few areas,

Table A4.12   Welfare analysis for model A2.3 (consumer subsidy)

J	PGN3 (1000¥)	CGN3 (1000¥)	BCS3 (1000¥)	NLS3 (1000¥)	S31 (00t)	D31 (00t)	P31 (¢/kg)
Beijing	0	1361090	1031277	329812	28087	11042	−5.72
Shanxi	18582	412770	461126	−29772	0	3628	−5.24
In.Mong.	−55874	426760	283651	87235	0	3026	−5.72
Liaoning	−116645	742383	550960	74777	0	5558	−5.72
Jilin	−185170	500554	624728	−309343	0	3464	−5.72
Heilongjiang	−120500	669890	809841	−260450	0	4995	−5.72
Shanghai	−1844	262343	128302	132196	−0	1819	−5.67
Jiangsu	0	1433150	1596864	−163713	22323	9295	−5.67
Zhejiang	294759	630079	966371	−41531	0	4130	−3.89
Anhui	0	1103049	1125121	−22071	7744	7744	−5.70
Fujian	139345	360952	461929	38369	0	2780	−3.89
Jiangxi	269389	544605	886669	−72674	0	3546	−3.89
Shangdong	0	1471341	1504408	−33067	11582	11582	−5.71
Henan	0	1471871	1626507	−154635	56805	11951	−5.88
Hubei	349017	790673	1260113	−120422	0	5184	−3.99
Hunan	269735	917605	1283529	−96188	0	5757	−3.89
Guangdong	176144	825001	893824	107321	0	6402	−3.89
Guangxi	90341	485624	545032	30933	0	3969	−3.89
Sichuan	984156	984804	2036456	−67496	0	7033	−2.62
Guizhou	57413	342134	321547	78000	0	3041	−3.89
Yunnan	243795	336851	517021	63625	0	2487	−2.62
Shaanxi	35227	510094	568578	−23256	0	4145	−5.24
Gansu	31167	372411	319485	84093	0	2820	−5.24
Qinghai	6510	100586	59990	47105	0	562	−5.24
Ningxia	9632	76578	87939	−1728	0	572	−5.24
Xinjiang	0	288881	296212	−7330	2217	2217	−6.18
$\sum$	2495184	17422092	20247489	−330212	128760	128760	—

Notes:
$PGN3_i = r_i(Q3_i - Q1_i)S3_i,$
$CGN3_i = \frac{1}{2}(D3_i + D1_i + 2D0_i)(P1_i - P3_i),$
$BCS3_i = (P_{si} - P_{ci})S3i,$
$NLS3_i = PGN3_i + CGN3_i - BCS3_i,$
$S31_i = S3_i - S1_i,$
$D31_i = D3_i - D1_i,$
$P31_i = P3_i - P1_i.$

PGN3, CGN3, BCS3 and NLS3 are respectively the regional vectors for producers' gains, consumers' gains, budgetary costs and the efficiency losses in model A2.3 as opposed to model A2.1.
S31, D31, P31 are respectively the regional differential vectors of supply, demand and price between model A2.3 and model A2.1.

i.e. Beijing, Jiangsu, Anhui, Shangdong, Henan, and Xinjiang. All these regions must be those with medium-level costs of production. Grain supplies in those regions with the lowest costs of production, cannot be increased as a result of a production subsidy because they are restricted by land constraints.

A4.2.4 Production subsidies and quotas (model A2.4)

Model A2.4 is built to test the combined effects of two policy instruments: a production subsidy (consumer subsidy) and a production quota.

(1) Welfare effects

In general, consumers in all regions will benefit more from this policy scenario than from a single policy instrument of either a production quota or a production subsidy. Let $CGN4_i$ denote the consumers' gain in region i, from the results of model A2.4 as opposed to the results of model A2.1, then

$$CGN4_i = \tfrac{1}{2}(D1_i + D4_i + 2D0_i)(P1_i - P4_i) \quad \text{for all } i \quad (A4.12)$$

Supply regions are divided into three groups according to Table A4.4 in order to calculate the regional gains and losses to producers:

(1) Group 1: $Q4_i > 0$ and $Q1_i > 0$. This implies that $r_iS1_i = r_iS4_i = L_i$.
(2) Group 2: $Q1_i = 0$ and $E4_i > 0$. This implies that $r_iS1_i \leq L_i$ and $r_iS4_i = I_i$.
(3) Group 3: $Q1_i = E4_i = 0$. This implies that $r_iS1_i < L_i$ and $I_i \leq r_iS4_i \leq L_i$.

The common formula can be written as:

$$PGN4_i = r_i(Q4_i - Q1_i - E4_i)S4_i \quad \text{for all } i \quad (A4.13)$$

If $BCS4_i$ denotes the budgetary cost in region i, then

$$BCS4_i = (P_{si} - P_{ci})S4_i \quad \text{for all } i \quad (A4.14)$$

Let $NLS4_i$ denote the net welfare loss to region i, then

$$NLS4_i = PGN4_i + CGN4_i - BCS4_i \quad \text{for all } i \quad (A4.15)$$

All the values of $PGN4_i$, $CGN4_i$, $BCS4_i$ and $NLS4_i$ for all i are presented in columns 2 to 5 of Table A4.13.

The total gains to consumers and total costs to the budget are substantial, about 20.25 billion yuan and 20.36 billion yuan respectively. The total loss to producers is 362 million yuan and the total efficiency loss is 471 million yuan.

(2) Impact on the production pattern

The differentials of supplies, demands and prices between the results of model A2.1 and model A2.4 are presented as $S41_i$, $D41_i$ and $P41_i$ for all i in the last three columns of Table A4.13. The introduction of a production quota into model A2.3 reinforces the effects of consumer subsidies on supplies, demands and prices.

Total supply and demand increase by almost 15 million tons, or 4.07 per cent. As for the pattern of production, the effect is similar to that of a production subsidy in model A2.3. The increase of supply is still concentrated in those six regions as in the above sub-section.

A4.2.5   Production tax (model A2.5)

As discussed in the previous chapter, if the producers are charged with a tax for each unit of output they produce, the effect on production, consumption and prices is the same as that by increasing the per unit production cost as implied in model A2.5. This sub-section analyses the effect of this policy scenario on the welfare of consumers, producers and the budget by comparing the results of models A2.5 and A2.1.

(1) The welfare effects

Since production costs increase as a result of a production tax, equilibrium prices are expected to increase in all regions. Consequently, regional demands for grain will decrease and consumer surplus be reduced. Let $CLS5_i$ denote the loss to consumers in region i, then

$$CLS5_i = \tfrac{1}{2}(D1_i + D5_i + 2D0_i)(P5_i - P1_i) \qquad \text{for all } i \qquad (A4.16)$$

On the supply side, the majority of regions are expected to lose although some regions may gain. According to Table A2.4, supply regions are classified into three groups in order to calculate the losses (or gains) to producers for all regions:

Table A4.13   Welfare analysis for model A2.4 (consumer subsidy and production quotas)

j	PGN4 (000¥)	CGN4 (000¥)	BCS4 (000¥)	NLS4 (000¥)	S41 (00T)	D41 (00T)	P41 (¢/kg)
Beijing	−288248	1711228	1153034	269946	49374	13801	−7.15
Shanxi	−66707	494515	461126	−33317	0	4327	−6.25
In.Mong.	−140483	536153	283651	112019	0	3782	−7.15
Liaoning	−283910	932945	550960	98074	0	6947	−7.15
Jilin	−388048	628788	624728	−383987	0	4330	−7.15
Heilongjiang	−352574	841826	809841	−320589	0	6244	−7.15
Shanghai	−17350	294494	128302	148840	−0	2035	−6.35
Jiangsu	−194400	1608542	1634916	−220774	29034	10402	−6.35
Zhejiang	220814	700713	966371	−44843	0	4585	−4.32
Anhui	0	1103049	1125121	−22071	7744	7744	−5.70
Fujian	103993	401514	461929	43579	0	3086	−4.32
Jiangxi	201045	605651	886669	−79972	0	3937	−4.32
Shangdong	−488266	1932843	1601700	−157123	28621	15110	−7.45
Henan	−158271	1633737	1483643	−8177	32508	13230	−6.51
Hubei	250209	877116	1260113	−132788	0	5741	−4.42
Hunan	157815	1020410	1283529	−105303	0	6393	−4.32
Guangdong	96775	917723	893824	120674	0	7109	−4.32
Guangxi	40330	540251	545032	35550	0	4408	−4.32
Sichuan	811343	1148190	2036456	−76923	0	8186	−3.05
Guizhou	28365	380677	321547	87495	0	3376	−4.32
Yunnan	199419	392757	517021	75155	0	2894	−3.05
Shaanxi	−67983	610909	568578	−25652	0	4944	−6.25
Gansu	−24673	445890	319485	101731	0	3363	−6.25
Qinghai	−3797	120312	59990	56524	0	671	−6.25
Ningxia	2534	83743	87939	−1661	0	625	−5.72
Xinjiang	0	288881	296212	−7330	2217	2217	−6.18
$\sum$	−362068	20252857	20361717	−470928	149498	149498	—

Notes:
$PGN4_i = r_i(Q4_i - Q1_i - E4_i)S4_i$,
$CGN4_i = \frac{1}{2}(D4_i + D1_i + 2D0_i)(P1_i - P4_i)$,
$BCS4_i = (P_{si} - Pci)S4_i$,
$NLS4_i = PGN4_i + CGN4_i - BCS4_i$,
$S41_i = S4_i - S1_i$,
$D41_i = D4_i - D1_i$,
$P41_i = P4_i - P1_i$.

PGN4, CGN4, BCS4 and NLS4 are respectively the regional vectors for producers' gains, consumers' gains, budgetary costs and the net losses from model A2.4 compared with model A2.1.
S41, D41, P41 are respectively the regional differential vectors of supply, demand and price between model A2.4 and model A2.1.

Group 1: $Q1_i > 0$ and $Q5_i > 0$. This implies that $r_iS1_i = r_iS5_i = L_i$.
Group 2: $Q1_i = Q5_i = 0$, or $P5_i = P_{ri}$ and $P1_i = P_{si}$. This implies that $r_iS1_i < L_i$ and $r_iS5_i < L_i$.
Group 3: $Q1_i > 0$ and $Q5_i = 0$. This implies that $r_iS1_i = L_i$ and $r_iS5_i \le L_i$.

A common equation for all regions is:

$$PLS5_i = r_i(Q1_i - Q5_i)S5_i \qquad \text{for all } i \qquad (A4.17)$$

Let $BGN5_i$ denote the budgetary gain in region i, then

$$BGN5_i = (P_{ri} - P_{si})S5_i \qquad \text{for all } i \qquad (A4.18)$$

Where P_{ri} is the new production cost with a unit production tax in region i. $P_{ri} = APP3_i$ (Chapter A3).

The net social welfare loss in region i is the difference between the budgetary gain and the sum of the producers' loss plus the consumers' loss in the same region. Let $NLS5_i$ denote the net loss in region i, then

$$NLS5_i = BGN5_i - PLS5_i - CLS5_i \qquad \text{for all } i \qquad (A4.19)$$

Producer and consumer losses, budgetary gains and efficiency losses for all regions are presented in columns 2 to 5 of Table A4.14. In many regions producers suffer heavy losses because of a production tax. In a few regions, producers benefit. This is because production taxation not only depresses production but also pushes up prices. Thus, producers in some regions (usually the low-cost and surplus regions) may gain more from higher prices than what they lose due to lower production as a result of taxation. The losses to consumers and budgetary gains in all regions are very large. The total efficiency loss to the society is also significant (more than 1.6 billion yuan) although some regions, such as regions Shanxi, Jilin, and Heilongjiang, have net social welfare gains.

(2) Impact on production patterns

The differentials of supplies, demands and prices between the results of models A2.1 and A2.5 are presented in the last three columns of Table A4.14. Total supply and demand decrease by more than 27.6 million tons or 7.53 per cent. Although optimum demands decline in all regions, optimum supplies decline in only a few regions. It is obvious that those regions which reduce their supplies are those with relatively

Results and Policy Implications 293

Table A4.14   Welfare analysis for model A2.5 (production taxation)

j	PLS5 (000¥)	CLS5 (000¥)	BGN5 (000¥)	NLS5 (000¥)	S51 (00T)	D51 (00T)	P51 (¢/kg)
Beijing	0	2530586	1016738	−1513848	−63329	−22065	11.43
Shanxi	83384	729354	921408	108669	0	−6869	9.92
In.Mong.	−69177	751831	567302	−115350	0	−5661	10.70
Liaoning	−188393	1342944	1103091	−51460	0	−10728	11.04
Jilin	−369834	940868	1249455	678421	0	−6923	11.43
Heilongjiang	−238768	1253010	1619681	605439	0	−9982	11.43
Shanghai	7582	473682	256835	−224429	−0	−3484	10.86
Jiangsu	0	2708191	2602785	−105406	−29788	−18590	11.34
Zhejiang	55076	1673947	1870316	141293	−5583	−11541	10.86
Anhui	0	2073664	1985377	−88287	−15489	−15489	11.40
Fujian	63633	911035	923030	−51638	0	−7432	10.39
Jiangxi	127006	1386945	1452654	−61297	−28968	−9479	10.39
Shangdong	0	2744278	2609725	−134553	−23164	−23164	11.42
Henan	0	2730720	1363945	−1366775	−103731	−23882	11.75
Hubei	−172118	2208201	2515652	479568	0	−15279	11.75
Hunan	117556	2122645	2569667	329465	0	−13911	9.39
Guangdong	49431	1888764	1785800	−152396	0	−15469	9.39
Guangxi	405998	701219	1090066	−17151	0	−5975	5.85
Sichuan	1098388	2681878	4076941	296674	0	−19892	7.40
Guizhou	141826	618168	643094	−116899	0	−5790	7.40
Yunnan	262970	916180	1034042	−145108	0	−7034	7.40
Shaanxi	237940	807773	1136137	90423	0	−6962	8.80
Gansu	210809	525671	639520	−96961	0	−4187	7.78
Qinghai	40387	143903	119879	−64412	0	−835	7.78
Ningxia	18960	144680	155368	−8272	−1727	−1149	10.52
Xinjiang	0	537887	509524	−28363	−4445	−4445	12.39
$\sum$	1882656	35548024	35818032	−1612648	276224	−276224	—

Notes:
$PLS5_i = r_i(Q1_i - Q5_i)S1_i,$
$CLS5_i = \frac{1}{2}(D5_i + D1_i + 2D0_i)(P5_i - P1_i),$
$BGN5_i = (P_{ri} - P_{si})S5_i,$
$NLS5_i = BGN5_i - PLS5_i - CLS5_i,$
$S51_i = S5_i - S1_i,$
$D51_i = D5_i - D1_i,$
$P51_i = P5_i - P1_i.$

PLS5, CLS5, BGN5 and *NLS5* are respectively the regional vectors of producers' losses, consumers' losses, budgetary gains and the net losses between model A2.5 and model A2.1.
S51, D51, P51 are respectively the regional differential vectors of supply, demand and price between model A2.5 and model A2.1.

high production costs. Therefore, the introduction of a producers' tax, which is proportional to the regional production costs for model 4.1 for each region, actually makes the grain supply more concentrated in the low-cost regions. Consequently, the loss in social welfare can not be explained by the loss in allocative efficiency. Rather, it is explained by the lack of incentive to producers in the high-cost regions to supply more grain to the market. This leads to a substantial reduction in consumer demand. And it is the massive losses in consumer's surpluses that explain the large social welfare losses.

5.2.6   Production taxation and quotas (model A2.6)

The combined effects of two policy instruments, i.e. production taxation and production quota will, is analysed in this sub-section based on the results of model A2.6.

(1) Welfare effects

Usually, a production taxation tends to reduce supply and demand and increase market prices of grain whereas a production quota has the opposite effect. The welfare effects of this policy scenario is more complicated than those of a single policy. These are discussed below.

Let $CLS6_i$ denote the consumer welfare loss in region i from model A2.6 as opposed to the results from model A2.1. The equation for $CLS6_i$ has the same form of (A4.16).

$$CLS6_i = \tfrac{1}{2}(D1_i + D6_i + 2D0_i)(P6_i - P1_i) \qquad \text{for all } i \qquad (A4.20)$$

To calculate the welfare effects on producers, supply regions are divided into four groups according to the results in Table A4.4:

(1) Group 1: $Q1_i > 0$ and $Q6_i > 0$.
    This implies that $r_iS1_i = r_i > S6_i = L_i$.
(2) Group 2: $Q1_i > 0$ and $E6_i > 0$.
This implies that $r_iS1_i = L_i$ and $r_iS6_i = I_i$.
(3) Group 3: $Q1_i = 0$ and E6_i > 0.
    This implies that $r_iS1_i < L_i$ and $r_iS6_i = I_i$.
(4) Group 4: $Q1_i > 0$ and $E6_i = Q6_i = 0$.
    This implies that $r_iS1_i = L_i$, and $I_i \leq r_iS6_i \leq L_i$.

Let $PLS6_i$ denote the welfare loss to producers in region i. It can be calculated by the following equation:

$$PLS6_i = r_i(Q1_i + E6_i - Q6_i)S6_i$$
$$+ r_iQ1_i(S1_i - S6_i) \quad \text{for all } i \qquad \text{(A4.21)}$$

The budgetary gain for region i denoted by $BGN6_i$ can be expressed as:

$$BGN6_i = (P_{ri} - P_{si})S6_i \quad \text{for all } i \qquad \text{(A4.22)}$$

The efficiency loss in region i denoted by $NLS6_i$ is the differential between the budgetary gain and the sum of the producer loss plus the consumer loss in the same region, thus

$$NLS6_i = BGN6_i - PLS6_i - CLS6_i \quad \text{for all } i \qquad \text{(A4.23)}$$

The values of producer losses, consumer losses, budgetary gains and the efficiency losses for all regions are presented in columns 2 to 5 of Table A4.15. Producers in every region suffer a heavy loss. The total loss to producers is about 33 billion yuan. Consumers also suffer a significant loss. The total loss to consumers is more than 6 billion yuan. Compared with the losses to consumers in model A2.5, however, consumers are much better off in model A2.6. The government has a huge budgetary gain of more than 38.6 billion yuan. The majority of regions in model A2.6 have some net welfare losses while a few regions (Jiangsu, Anhui, Henan, and Hunan) have a gain. The total efficiency loss to the society is about 723 million yuan, which is much smaller than that from model A2.5. The contrast between the efficiency losses of models A2.5 and A2.6 has some interesting policy implications as will be discussed later.

(2) The impact upon production patterns

The differentials of supplies, demands and prices between models A2.1 and A2.6 are presented in the last three columns of Table A4.15. Grain demands in all regions except Shangdong decrease under this policy scenario. Total supply and demand decrease by 4.7 million tons. This is rather insignificant when it is compared with that of model A2.5. It implies that the introduction of a production quota has a remarkable counter effect on the sharp decline in supply induced by production taxation. Column 5 (S61) shows that in some regions, such as Beijing,

Table A4.15   Welfare analysis for model A2.6 (production taxation and quotas)

j	PLS6 (000¥)	CLS6 (000¥)	BGN6 (000¥)	NLS6 (000¥)	S61 (00T)	D61 (00T)	P61 (¢/kg)
Beijing	2225637	90501	2306068	−10069	49374	−752	0.39
Shanxi	704370	130586	797365	−37592	−11369	−1184	1.71
In.Mong.	543841	28466	567302	−5005	0	−206	0.39
Liaoning	1025177	49459	954598	−120037	−15746	−379	0.39
Jilin	1193994	33407	1163624	−63777	−9731	−236	0.39
Heilongjiang	1397096	44633	1401639	−40091	−21891	−340	0.39
Shanghai	203537	54649	222237	−35948	−3091	−388	1.21
Jiangsu	2920352	299015	3269832	50464	29034	−1984	1.21
Zhejiang	1118635	592186	1672550	−38270	−23273	−3985	3.75
Anhui	1947257	235994	2187189	3937	2213	−1699	1.25
Fujian	592952	295943	798769	−90126	−11144	−2346	3.28
Jiangxi	1104832	448480	1537339	−15973	−21393	−2993	3.28
Shangdong	3208764	−5039	3200594	−3130	28621	39	−0.02
Henan	2828185	131648	2964764	4930	32508	−1099	0.54
Hubei	1736509	475247	2176986	−34771	−30787	−3185	2.45
Hunan	1901834	589358	2569667	78474	0	−3777	2.55
Guangdong	1313074	527290	1785800	−54564	0	−4201	2.55
Guangxi	791562	309956	1090066	−11452	0	−2604	2.55
Sichuan	3050592	940784	3939512	−51864	−13566	−6854	2.55
Guizhou	470588	217855	643094	−45349	0	−1995	2.55
Yunnan	768640	321588	1034042	−56186	0	−2424	2.55
Shaanxi	901273	142960	983190	−61043	−13717	−1195	1.51
Gansu	526585	81824	553434	−54975	−7414	−635	1.18
Qinghai	98948	22244	103739	−17452	−1373	−126	1.18
Ningxia	131185	25895	152072	−5008	−2006	−198	1.82
Xinjiang	262034	282557	536347	−8244	−2277	−2277	6.35
$\sum$	32967450	6367486	38611819	−723117	−47023	−47023	—

Notes:

$PLS6_i = r_i(Q1_i + E6_i − Q6_i)S6_i + r_iQ1_i(S1_i − S6_i)$,

$CLS6_i = \frac{1}{2}(D6_i + D1_i + 2D0_i)(P6_i − P1_i)$,

$BGN6_i = (P_{ri} − P_{si})S6_i$,

$NLS6_i = BGN6_i − PLS6_i − CLS6_i$,

$S61_i = S6_i − S1_i$,

$D61_i = D6_i = D1_i$,

$P61_i = P6_i − P1_i$.

PLS6, CLS6, BGN6 and NLS6 are respectively the regional vectors of producers' losses, consumers' losses, budgetary gains and efficiency losses of model A2.6 compared with model A2.1.

S61, D61, P61 are respectively the regional differential vectors of supply, demand and price between model A2.6 and model A2.1.

Jiangsu, Anhui, Shangdong and Henan, grain output declines sharply in model A2.5 but greatly increases in model A2.6. The opposite changes of grain production in these regions from model A2.5 to model A2.6 suggest that producers in these regions suffer enormously under the combination of production taxation and quota. As for the relatively low-cost regions, such as In.Mon, Hunan, Guangdong, Guangxi, Guizhou and Yunnan, grain supply in model A2.6 is roughly the same as in model A2.1. These regions still provide their maximum supplies. The production decline is concentrated in the 'medium-cost' regions such as regions Shanxi, Liaoning, etc.

4.3   POLICY IMPLICATIONS

From the analysis in section 2, it is obvious that any policy instrument or any combination of policy instruments have two distinct effects on the grain sector; a welfare redistribution effect and a production redistribution effect.

As for welfare redistribution, there are two relationships which deserve mentioning here:

(1) welfare redistribution among producers, consumers and the taxpayers (the government budget).
(2) welfare redistribution among regions.

On the production side, the 'basic' or 'efficient' production pattern is usually distorted by any intervention of the government. The pattern is usually altered in such a way that more grain is produced in those 'inefficient' regions at a higher cost than it would be in the absence of intervention. The distorted production pattern, of course, can be used to explain the redistribution of welfare between different groups of people and between different regions. It can also be used to explain the net social welfare or efficiency losses.

As mentioned many times, government intervention always incurs some efficiency loss to society resulting in welfare redistribution which itself may also be undesirable. One critical question is why the government wants to intervene? This is already discussed in Chapter 2. One related question is what kind of intervention can achieve government's goals most effectively and efficiently.

We now discuss the second question by comparing the outcomes of different policy scenarios of the previous section.

A4.3.1 Production quotas

A production quota has two positive effects: (1) it raises production and suppresses price; and (2) it increases consumer surplus. The negative effect is that it incurs substantial losses to producers.

A4.3.2 Consumer subsidies

Consumer subsidies have the following positive effects: (1) lower price and more output; and (2) greater consumer and producer surplus. The negative effect is also obvious: the government suffers a heavy budgetary loss.

A4.3.3 Consumer subsidies and production quotas

Compared with the outcomes of consumer subsidies alone, consumers under this scenario gain much more benefit and the government incurs slightly more cost to its budget. As for producers, the negative effect of a production quota outweighs the positive effect of subsidies, but the total loss to producers is not very significant. The efficiency loss also increases slightly.

A4.3.4 Production taxation

There are three negative effects in this case: (1) the prices are much higher and consumers suffer substantial losses; (2) it incurs the greatest efficiency loss among all the policy scenarios in this study; (3) total grain output is substantially reduced and producers suffer heavy losses. The only positive effect is that the government has a sizeable budgetary gain.

A4.3.5 Production taxation and quotas

Since most effects of production quota and taxation are opposite, all the negative effects except that on producers induced by production taxation alone, are greatly reduced under this policy scenario. Compared with those of production taxation, the loss to consumers, the efficiency loss, and the loss in grain supply under this scenario are all greatly reduced. Market prices are also reduced by a great margin. The budgetary gain also increases because the more producers produce, the more they have to be taxed. The loss to producers is enormous. It is the greatest among all the policy scenarios in this study.

The above analyses demonstrate the costs and benefits of different policy scenarios for the decision-makers. According to these analyses, if the principal objectives of government are to increase total grain supply and bring down prices to favour consumers, a production quota is more effective than other policies, although it incurs significant losses to producers. However, if the government is more concerned with producers' income, but still wants to increase total grain production and subsidize consumers, the introduction of production subsidies or combination of production subsidies with production quotas will help achieve these goals, although these two polices incur substantial budgetary costs to the government. If the budgetary gain (or cost) is more important than the interests of producers and consumers, the imposition of production taxation is more appropriate. But by imposing production taxation, decision-makers must be aware that efficiency loss and the loss in production are substantial without the help of production quotas. The damage to producers, however, is very severe with the combination of production taxation and quotas.

In short, any government intervention will involve a trade-off. The choice of policies should be justified by their costs and benefits. A critical point is that decision-makers should be aware of the consequences of their decisions in order to achieve their goals effectively without too harmful an effect.

Notes

1 Agricultural Policies before Reform

1. Exceptionally low population growth of 0.8 per cent per year from 1957 to 1962, compared with an average of more than 2 per cent per year in the 1950s and 1960s, is recorded in Table 2.11.
2. The re-establishment of the three-level system in the commune was a fundamental diversion from the single commune management system practised in the Great Leap Forward movement. The former is far superior in the sense that it promotes greater accountability and individual incentives. This same system continued to exist throughout the Cultural Revolution. In retrospect, the three-level system might have prevented China's agriculture from being totally collapsed for a long period of time and made the post-Mao reforms possible and successful. This is a great contrast to the Soviet system which was characterized by a much higher level of collectivization.
3. Private plots are pieces of communal land allocated to be used by individuals. Farmers can produce whatever they like on the plots and retain all the outputs. The term 'private' means the 'usufruct right and the ownership of outputs', but not 'entitlement'. Private plots generally constituted 5–10 per cent of total communal land but they were totally abolished during the Great Leap Forward movement. During the Cultural Revolution, these private plots were reduced to less than 5 per cent of total communal land. In some regions and some years they were totally abolished.
4. The degrees of state control on these activities varied significantly in different years and across regions, depending largely on the momentum of political struggle and the attitudes of local bureaucracies. At the peak of the Cultural Revolution, farmers were not allowed to produce any cash crops at all in some regions and poultry production was limited to a minimum. For example, the number of chickens, ducks and geese was limited to less than five, and the number of pigs less than two for any individual household in eastern Guangdong.

2 Agricultural Reforms

1. Jieyang county has recently been promoted to become a prefectural city, parallel to Shantou and Chaoan cities. This means that the former Shantou Prefecture has been divided into three equally important prefectural cities, reflecting the rapid development of the economy and urbanization in eastern Guangdong.
2. Not necessarily the seeds with the greatest yield potential. Seeds with high yield potential have to be supported by adequate fertilizers and pesticides. In the Cultural Revolution, villagers were forced by the commune cadres to adopt HYVs with the highest yield potential. Those HYVs were

300

vulnerable to rice insects and diseases, especially before harvest. In some years, up to half of the total harvest was destroyed by insects and diseases because there were not enough insecticides and pesticides to control them. This was one of the many reasons why farmers were very disappointed by the commune system in which they could not make their own decisions about production, even in choosing the right variety of rice.

3. During the commune period, harvest was mainly done by a group of adults in the production team. Each team had about 20–30 adults and 6.7 hectares of paddy. It usually took 12–15 days to complete the harvest and these adults (with the help of students on school holidays) had to work more than 10 hours in the field. However, the harvest might take less than 4 days if they were divided into individual households as is now the case under the household responsibility system.

4. 1 *jin* is equal to 0.5kg. 1 *mu* is equal to 1/15 hectares. In my village, rice is grown twice a year. If one harvest yields 1200–1400 *jin* of unprocessed rice per *mu*, two harvests should yield twice as much. Thus, annual yield per mu should be 2400–2800 *jin* of unprocessed rice. This is equivalent to 18–21 tons per hectare of unprocessed rice, or 12.6–14.7 tons per hectare of processed rice.

3 Policy Changes during Economic Reforms

1. For example, Lin (1988), McMillan *et al.* (1989) and Fan (1991).
2. In the early 1980s, when Zhao Ziyang was the prime minister, he stated that two factors could be relied upon to stimulate agricultural growth: 'the first is correct policies, and the second is sciences' (*yi kao zheng che, er kao ke xie*). It seems that 'a third element of investment' was missing. It is true that reforming policies and sciences are two major stimuli of agricultural growth during the reforms, but the lack of state investment has greatly undermined the long-term prospects of sustainable growth (see Chapter 4).
3. Lardy (1983) considers increased agricultural investment as one of the major policies adopted during economic reforms to accelerate agricultural growth. Many other authors also shared the same view (see Yao, 1987). This judgement is biased purely due to the lack of detailed investment data in the early 1980s. As more data appeared in the press, it appeared that investment in agriculture and rural infrastructure had suffered enormous cuts during the 1980s. Thus, the success of agricultural reforms between 1978 and 1984 has to be attributed to factors other than increased investment in agriculture.
4. Rural township and village enterprises (RTVEs) refer to those established and run individually and collectively by the farmers in rural towns (formerly the commune towns) and villages (formerly the production brigades or villages). Their activities range from industries, construction, transportation and various services activities.

   Detailed studies on China's rural industry are cited in Bhalla (1991), and Byrd and Lin (1990).
5. Detailed discussion of state policies on rural non-farm enterprises, see Guan Ze-wen (1991).

4   Problems and Prospects of Grain Production

1. Although direct expenditure on grains is much lower (about 12 per cent of total expenditures), most food items are derived from grains.
2. The state council agreed in 1979 that the share of agricultural investment in total state investment would be raised up to 18 per cent for the period of 1980 to 1984, from 14 per cent in 1979, and 12 per cent in the pre-reform period. This is quoted by Lardy in 1983 (Lardy, 1983). However, in implementation, the share was curtailed to about 5 per cent. This reflects the legacy of how agriculture is neglected and how investment decision is *ad hoc* and myopic. The heart of the matter lies in a strong tendency for politicians to achieve *quick and high* economic growth through expanding industries without thoughtful consideration of long-term efficiency and the need to balance the structure of the economy.
3. The changes were primarily to cut the state budgetary costs by transferring more consumer subsidies to be borne by farmers although the official intention was to deepen price reforms and to boost farmers' incomes.
4. A single mixed procurement price is a weighted average price of the former basic procurement price paid to quota delivery and the former premium price paid to above-quota delivery.
5. To stimulate grain production, a positive step is to raise the price first. If the procurement price is not as high as the free price, cutting the quota will inevitably lead to a reduction in output and increase inefficiency in resource allocation (see analyses in Chapter 7, and the Appendix).
6. Per capita cultivated land was 0.34 hectares in the rest of the world in 1984 (Chen and Buckwell, 1991, Table 1.2).
7. In 1989, for instance, the average free market price of grain was 1.03 yuan/kg, while the contract procurement price was 0.54 yuan/kg and the mixed average procurement price was 0.61 yuan/kg (Research Centre of Economic Policy, MOA, 1991, p. 180).
    The mixed procurement price is the weighted average of the contract procurement price and the state negotiated price. The weights used to calculate the mixed average price are different across regions, thus the mixed average price varies in different areas. Between 1979 and 1985, there was a two-tier price system: the contract (quota) price with which farmers were forced to sell a quota to the state, and the negotiated price which was the contract price plus a premium. After 1985, the two-tier price system was replaced by the mixed average procurement price (see Section 4.1).
8. Farmers have difficulty in selling grains to the state at contract prices. They are either denied for delivery or not paid promptly; as the Chinese put it, marketing agencies 'writing an empty cheque' to farmers.

5   Agriculture's Role in the Chinese Economy

1. The share of agricultural output in total agricultural and industrial output increased for the first time (and probably the only time) in modern Chinese history. This provides indirect evidence showing that agriculture was

excessively undermined by government policies in the pre-reform era (for a detailed discussion, see Chapter 2, and Table 2.9).

2. The Chinese experiences are in sharp contrast to those observed in the former Soviet Union, some other Eastern European countries, and many African states. Throughout the 1980s, these countries suffered a great deal from super-high inflation, largely induced by acute food shortage. The experiences of the former Soviet Union have shown how a failure to achieve sustainable agricultural growth can lead to a total collapse of the whole economy.

3. On average, direct agricultural tax is about 6 billion yuan per year, or 3 per cent of total agricultural output value (Economic and Policy Research Centre, MOA, 1991, vol. 2. p. 438).

4. Government intervention in pricing results in lower prices paid for agricultural goods and higher prices of industrial goods charged to farmers than their respective 'real market values'. The divergence curves of regulated prices away from their real market values look like the two blades of scissors when they are open. The distance of the two blades from the central axis (implicitly referred to the real market values) measures the degree of divergence of regulated prices from their respective real values. The 'scissors difference', therefore, refers to the sum of the difference between the regulated prices and their real market values of agricultural goods, and that of industrial goods sold to the rural areas. This scissors difference can be illustrated by the diagram below.

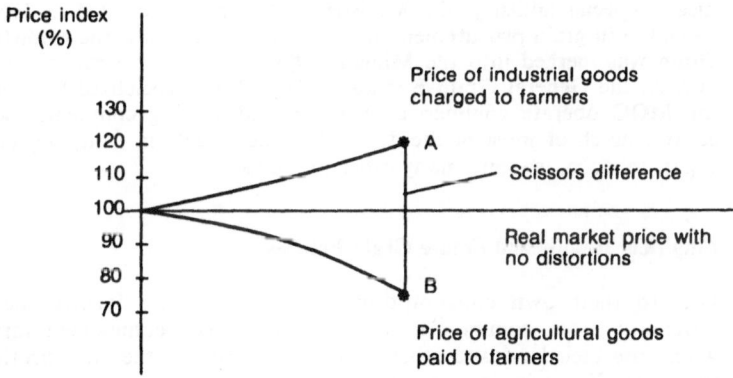

5. Although some authors have highly praised the outcomes of economic reforms in the state industrial sector (see McMillan and Naughton, 1992), my view is that reforms in this sector have not been satisfactory. As in the pre-reform period, impressive output growth of this sector in the reform period has been accompanied by declining profitability and increasing losses.

304 *Notes*

6 Policy Objectives, Instruments and Mechanisms

1. In many Asian and African countries, the primary objective of food subsidies is to support the vulnerable groups. In practice, however, few countries have been successful in targeting subsidies, either because it is difficult to define those who are in need of support and those who are not, or because the cost of administration is too high. As a result, most countries have opted to have a universal subsidy programme which benefits all the population or the whole of a particular segment of the population (such as all the urban consumers).
2. In practice, all the other activities were sacrificed for the production of foodgrains (see Chapter 1).
3. In most Eastern and Southern African countries, the marketing of each major agricultural product (e.g., maize, cotton, coffee) was almost entirely controlled by a separate state marketing parastatal before market reforms (Beynon, Jones and Yao, 1992). The parastatal exercised a whole range of policy instruments. The wave of market liberalization led by the World Bank and the IMF has either eliminated many marketing boards and/or reduced their power to control since the 1980s. However, government intervention on food marketing in Africa is still very important and affects strongly the performance of food production in that part of the world (Yao and Hay, 1991). Thus the discussion of policy instruments and their implementation mechanism on China will be very relevant to Africa and other developing countries.
4. As grain and vegetable oil were the most important controlled commodities, a special ministry, the Ministry of Grain, was created in 1956 to specialize in grain procurement and distribution. In 1982, the Ministry of Grain was merged into the Ministry of Commerce as a sub-ministerial bureau, the General Grain Bureau (GGB). Other specialized bureaux of the MOC operate commercial agencies and trading companies which control much or most of the domestic trade in cotton, fruit, vegetables, sugar, tobacco, tea, and many other products.

7 Empirical Results and Future Grain Policies

1. Due to their own consumption needs, subsistence farmers behave differently from commercial farmers. In general, commercial farmers would use their land and other resources to produce the most profitable products. For security reasons, subsistence farmers usually have to produce enough food for their own consumption first before they can produce marketable grains or cash crops. Thus profit maximization may not be their primary objective. Although most grain farmers in China are not mere subsistence producers because they sell a significant proportion of their output to the market, they still have to produce enough food for their own consumption. In a purely technical sense, the value attached to their own consumption could be higher than the farm-gate price.
2. The latest data reveals that the total output value of the RTVEs increased by 70 per cent in the first five months of 1993 over the same period of 1992.

3. A typical example is a recent battle between the two southern provinces of Guangdong and Hunan on rice trading. Farmers in Guangdong, the richest and fastest growing region in China, found it much more profitable to produce vegetables, fruits and other cash crops than foodgrains, leaving Guangdong demanding more rice imports from the neighbouring rice-producing Hunan province. For financial reasons, the Hunan government decided to impose heavy duties on rice export to Guangdong. To avoid taxes, the Guangdong government bypassed the Hunan authority and directly purchased rice from the local farmers of Hunan at a price higher than the local official procurement price but lower than the exporting price to Guangdong. This angered the Hunan authority who then sent troops to stop rice exports from Hunan to Guangdong; Guangdong countered with its own mobilisation (*The Economist*, 26 June–2 July, 1993, p. 71).

4. Cross-subsidies from non-farm to farm activities have been widely observed in Jiangsu, Zhejiang, Guangdong, Shanghai and Beijing since 1985. This phenomenon reflects the huge potential for regional specialization of production. It also reflects the degree of market imperfection in China as illustrated by the Guangdong–Hunan conflict on rice trading.

A1 Geometric Analysis of Selected Policies

1. In Tanzania by 1980 crop trading authorities had worked up a combined deficit equal to 15 per cent of GDP. In Zimbabwe about 3.5 per cent of the government's budget went to making good the losses of marketing board (Harriss, 1990). In Zambia, total subsidies to the food and agricultural sector, of which 95 per cent directly or indirectly subsidized maize consumption, reached 18 per cent of government recurrent spending in 1980 (Pearce, 1990). In Egypt, the share of food subsidies in total public spending increased dramatically from less than 1 per cent in 1970 to about 17 per cent by 1974 (Alderman and Braun, 1984, p. 12). In the first half of the 1980s, the bill for Egyptian food subsidies was around US$2000 million annually (Pinstrup-Andersen, 1988, p. 13). In Sri Lanka, the cost of food subsidies was more than 15 per cent of government expenditures during the late 1970s although it was reduced to less than 3 per cent by 1985 due to targeting subsidies and falling real prices of food procured by the government (ibid). In Bangladesh, the share of total government budget was about 10 per cent excluding food aid and 3.4–8.5 per cent including food aid in the 1977–80 period (ibid, p. 223). The cost of food subsidies in Pakistan measured in terms of the share of total government expenditure was as high as 10.3 per cent in 1974 but declined to only 1.3 per cent in 1984 (ibid, p. 250, Table 17.1).

2. This implies that there are no administrative trading barriers or government interventions on regional and inter-regional trade.

3. In writing up this chapter, the author noticed that the Chinese government still used input supply at a fixed price as an incentive to encourage grain production and delivery to the state (*People's Daily*, 18 March 1993, p. 1). In recent years, the major advantage of guaranteed supply of farm inputs is probably not that of low price but that of guaranteed availability at a predictable cost. There was some scandalous behaviour in input marketing

in the late 1980s. Farmers suffered immensely due to the profit-seeking and monopolistic suppliers of farm inputs. Thus, the guaranteed supply of inputs at a fixed price is perhaps a means to free farmers from exploitation by the marketers.

4. In fact, all policy instruments trigger a set of trade-offs. These are not discussed in the analyses of other policy instruments. However, the trade-off analyses for input subsidy in this section are relevant for other policy analyses.

A2 Policy Models for the Chinese Grain Sector

1. The theoretical models of quadratic spatial equilibrium are discussed in Yao (Yao, 1989c, Chapter 3), Takayama and Judge (1971) and Rayner (1974). A literature survey of spatial equilibrium analysis is done by Weinschenck *et al.* (1969). Actual solution methods are developed by Enke (1951), Wolfe (1959) and Dantzig (1963).

2. In this book, matrix algebra of the mathematical models are entirely avoided. It is hoped that non-matrix algebra will be more friendly to the general readers. However, matrix algebra has many attractions. For those who prefer matrix notations, see Chapters 3 and 4 of Yao (1989c).

A3 Data Processing

1. The state procurement price was much lower than the free market price. Even the premium procurement price, which was 50 per cent higher than the basic procurement price, was lower than the free market price in most areas. Data for the free market prices in 1985 are not available. However, the State Administrative Bureau for Industry and Commerce conducted a price survey in October 1987. The survey covered 137 grain markets in twenty-six regions. According to the survey, the average free market price of grain was 63.1 yuan/100kg. Comparing with this and allowing for an inflation rate of about 10 per cent between 1985 and 1987, the average value of 50.05 yuan/100kg of APP_1 in Table A3.10 is not unrealistic. Therefore, the assumption made above is considered to be a reasonable one and the values of APP_1 are thought to be a reasonable approximation of the real demand prices of grain for 1985.

Bibliography

The following abbreviations are used in the references:

BR	*Beijing Review*
CAYB	*Chinese Agricultural Yearbook*
CCPCC	Chinese Communist Party Central Committee
CCYC	*Chien-chi Yen-chiu*
CKSHKH	Chung-kuo She-hui Ko-hsueh
CSYB	*China Statistical Yearbook*
FAO	Food and Agriculture Organization of the United Nations
GATT	General Agreement on Tariffs and Trade
GDP	Gross domestic product
GNP	Gross national product
HC	Hong-chi
IFPRI	International Food and Policy Research Institute
IMF	International Monetary Fund
JMJP	*People's Daily*
MOA	Ministry of Agriculture
SSB	State Statistical Bureau
ZGNCJJTJDQ	China Rural Statistical Compilation
ZGNYTJZL	*China Agricultural Statistical Data*

Agricultural Yearbook Compilation Commission (1981) *Chinese Agricultural Yearbook* (Agricultural Publishing House, Beijing).

Aird, John S. (1980) 'Reconstruction of an Official Data Model of the Population of China', unpublished manuscript, US Bureau of the Census.

Alderman, H. and Joachim von Braun (1984) *The Effects of the Egyptian Food Ration and Subsidy System on Income Distribution and Consumption*, IFPRI Research Report no. 45.

Alderman, H., Joachim von Braun and Sakr Ahmed Sakr (1982) *Egypt's Food Subsidy and Rationing System: A Description*, IFPRI Research Report, no. 34.

An Xi-Ji (1985) 'Pricing System Reform for Agricultural Products and Price Policy Adjustment in China 1979-84'. Paper presented to the 19th International Conference of Agricultural Economics held at Malaga, Spain, 26 August–4 September 1985.

Anderson K. and R. Tyers (1987) 'Economic Growth and Market Liberalization in China: Implication for Agricultural Trade', *The Development Economies*, June (2), Institute of Developing Economies, Tokyo.

Barankin, E. W. and R. Dorfman (1958). 'On Quadratic Programming', *University of California Publication: Statistics*, vol. 12, pp. 285–318.

Beijing Review (1985a) 'Communique on the 1984 National Economy', 25 March, Documents.

BR (1985b) 'Family-planning Policy Improves', 14 July.

BR (1985c) 'Economic Reform Brings Better Life', 22 July.

BR (1985d) 'Sixth Five Year Plan Succeeds', 16 September, p. 14.

BR (1988a) 'Grain Production: Today and Tomorrow', 8 February, p. 4.

BR (1988b) 'China's Agricultural Policies Challenged', 8–14 February, p. 6.

Beynon, J., S. Jones and S. Yao (1992) 'Market Reform and Private Trade in Eastern and Southern Africa', *Food Policy*, vol. 17, no. 6, pp. 399–408.

Bhalla, A. S. (1991) *Small and Medium Enterprises: Technology Policies and Options* (Greenwood Press, London).

Bhalla, A. S. (1992) *Uneven Development in the Third World* (Macmillan Press, London).

Butler, S. (1987) 'Grain Shortfall Will Raise Imports – Agriculture: Disposal of Farm Collectives Has Been a Problem', *Financial Times*, 18 December.

Byrd, William A. and Lin Qingsong (eds) (1990) *China's Rural Industry* (Oxford University Press, Oxford).

Caman H. F. and J. A. Moetzold (1971). 'The Regional Impact of Lamb Imports on Equilibrium Returns to Domestic Producers, 1967', *American Journal of Agricultural Economics*, 5, pp. 92–100.

Chen, L. Y. and A. Buckwell (1991) *Chinese Grain Economy and Policy* (CAB International).

Chinese Economic Yearbook (1982) (Statistical Publishing House, Beijing).

Colman, D. R. and F. I. Nixson (1986) *Economics of Change in Less Developed Countries* (Philip Allan Publishers, Oxford).

Colman, D. R. and T. Young (1990) *Principles of Agricultural Economics: Markets and Prices in Less Developed Countries* (Cambridge University Press, Cambridge).

Dantzig, G. B. (1963) *Linear Programming and Extensions* (Princeton University Press, Princeton, NJ).

Dodge, D. J. (1977) *Agricultural Policy and Performance in Zambia: History, Prospects and Proposals for Change* (Institute of International Studies, University of California at Berkeley).

Dodge, D. J. (1979) *Zambian Agricultural Policy and Market Policy*, mimeo (World Bank, Washington, DC).

Economic and Policy Research Centre, MOA (1991) *Rural China: Review of Policy Research*, vol. 2 (Reform Publishing House, Beijing).

The Economist (UK) (1986) 'International: Meanwhile, Back on the Farm', 18 January.

The Economist (UK) (1987) 'China's Economy Survey', 1 August.

The Economist (UK) (1993) 'Cut along the Dotted Lines', 26 June–2 July, p. 71.

Enke, S. (1951) 'Equilibrium Among Spatially Separated Markets: Solution By Electric Analogue', *Econometrica*, vol. 19, pp. 40–7.

Fan, S. (1991) 'Effects of Technical Change and Institutional Reform on Production Growth in Chinese Agriculture', *American Journal of Agricultural Economics*, vol. 73, no. 2, May.

FAO (1990) *Production Year Book and Fertilizer Year Book* (Food and Agriculture Organization of the United Nations, Rome).

Fei, J. C. and G. Ranis (1964) *Development of the Labour–Supply Economy: Theory and Policy* (Irwin, Homewood, Ill).

Ghatak, S. and K. Ingersent (1984) *Agriculture and Economic Development* (Wheatsheaf Books, Brighton).

Ghose, A. K. (1984) 'The New Development Strategy and Rural Reforms in Post-Mao China', in K. Griffin (ed.) *Institutional Reforms and Economic Development in the Chinese Countryside* (Macmillan, Hong Kong).

Grain Research Group of the Economy and Policy Research Centre for the Ministry of Agriculture, Animal Husbandry and Fishery (1988) 'Grain Shortages and Readjustment of Economic Policies', *Problems of Agricultural Economy*, no. 6, pp. 13–19.

Griffin, K. (ed.) (1984) *Institutional Reforms and Economic Development in the Chinese Countryside* (Macmillan, Hong Kong).

Guan Ze-wen (1991) 'Technological transformation of small and medium enterprises in China', in A. S. Bhalla (ed.) *Small and Medium Enterprises: Technology Policies and Options*, published on behalf of the United Nations Centre for Science and Technology for Development (Greenwood Press, London).

Guo Shutian (1991) *China's Cereals: Multi-angle Research and Thinking* (Agricultural Publishing House, Beijing).

Haggblade, S., P. Hazell and J. Brown (1989) 'Farm–non-farm linkages in rural Sub-Saharan Africa', *World Development*, vol. 17, no. 8, pp. 1173–1201.

Harriss, B. (1990) *The Deregulation of State Food Markets in Sub-Sahara Africa and South Asia*, Draft paper, Queen Elizabeth House, Oxford.

Hay, D. and S. Yao (1990) 'Production, Cost, Profit, Investment and Incentive Behaviour of the Chinese Manufacturing Sector, 1980–87: A Simultaneous and Non-linear Econometric System to Determine Production, Cost and Profits of Chinese Manufacturing Enterprises under Economic Reforms in the 1980s', Discussion Paper, Institute of Economics and Statistics, University of Oxford.

Hazell, P. B. R. (1984) 'Rural Growth Linkages and Rural Development Strategy', Paper prepared for the Fourth European Congress of Agricultural Economics, Kiel 3–7 September (IFPRI, Washington, DC).

Hazell, P. B. R., and A. Roell (1983) 'Rural Growth Linkages: Household Expenditure Patterns in Malaysia and Nigeria', Research Report No. 41. (International Food Policy Research Institute, Washington, DC).

Hicks, J. R. (1939) *Value and Capital* (Oxford University Press, Oxford).

Hirschman, A. O. (1958) *The Strategy of Economic Development* (New Haven, CT: Yale University Press).

Ishikawa, S. (1965) *National Income and Capital Formation: An Examination of Official Statistics* (Institute of Asian Economic Affairs, Tokyo).

Jan, G. P. (1986) 'The Responsibility System and Economic Impact on Rural China', *Asian Profile* 14(5), pp. 391–408.

Jiang Zheming, (1993) 'Correctly Control the Situation, Grip the Opportunity and Work Solidly to Ensure that the National Economy Develops Fast and Well'. *People's Daily (Overseas Edition)*, 1 June, p. 1.

Johnson, D. G. (1988) 'Economic Reforms in the People's Republic of China', *Economic Development and Cultural Change*, April, pp. 225–45.

Johnston, B. and P. Kilby (1975) *Agriculture and Structure Transformation: Economic Strategies in Late-developing Countries* (Oxford University Press, London).

Kojima, Reeitsu (1982) 'Prospects of New Chinese Agricultural Policy', *The Development Economies*, vol. 20, pp. 390–412 (Institute of Developing Economies, Tokyo).

Kojima, Reeitsu (1985) 'Characteristics of China's Present Agricultural Policy', *China Newsletter*, no. 58, pp. 2–10.

Kuhn, H. W. and A. W. Tucker (1951) 'Non Linear Programming', in *Proceedings of the Second Berkeley Symposium on Mathematical Statistics and Probability*, ed. J. Neyman, California University Press, pp. 481–92.

Kuznets, S. (1964) 'Economic Growth and the Contribution of Agriculture', in C. K. Eicher and L.W. Witt (eds), *Agriculture in Economic Development* (McGraw-Hill, New York).

Lardy, N. R. (1983) *Agriculture in China's Modern Economic Development* (Cambridge University Press, Cambridge).

Lewis, W. A. (1954) 'Economic Development with Limited Supplies of Labour', *Manchester Schools*, 22, pp. 139–91.

Lin, Justin Yifu (1988) 'The Household Responsibility System in China's Agricultural Reform: A Theoretical and Empirical Study', *Economic Development and Cultural Change*, April, pp. 199–224.

Markowitz, H. (1956) 'The Optimization of a Quadratic Function Subject to Linear Constraints', *Naval Research Logistics Quarterly*, no. 3, pp. 111–33.

Marshall, A. (1920) *Principles of Economics*, 8th edn (Macmillan, London; 1st edn, 1890)

McCalla, Alex F. and T. E. Josling (1985) *Agricultural Policies and World Markets* (Macmillan Publishing Company, New York).

McMillan J. and B. Naughton (1992) 'How to Reform a Planned Economy: Lessons from China', *Oxford Review of Economic Policy*, 8(1).

McMillan, J., J. Whalley and L. Zhu (1989) 'The Impact of China's Economic Reforms on Agricultural Productivity Growth', *Journal of Political Economics*, 97, pp. 781–807.

Mellor, J.W. (1976) *The Economics of Growth: A Strategy for India and the Developing World* (Cornell University Press, Ithaca, NY).

Mellor, J. and Uma J. Lele (1973) 'Growth Linkages of the New Food Grain Technologies', *Indian Journal of Agricultural Economics*, vol. 18, no. 1 (January–March), pp. 35–55.

Ministry of Agriculture Policy Research Office, (1980) *China's Basic Agricultural Situation* (Agricultural Publishing House, Peking).

Ministry of Food Research Office (1981) *A Discussion of Policy on Procurement and Sales of Cereals and Oils in Rural Areas* (Finance and Economic Publishing House, Beijing).

Ministry of Railways (1985) *A Handbook for Standing Charges for Railway Transportation* (Beijing).

Nolan, P. (1991) *Reforms and Social Development in China*, mimeo, Cambridge University.

Pearce, R. (1990) *Food Consumption and Adjustment in Zambia*, Working Paper no. 2, Food Studies Group, Oxford University.

People's Daily (1993), 18 March, p. 1.

Perkins, D. (1975) 'Growth and Changing Structure of China's Twentieth Century Economy', in D. Perkins (ed.) *China's Modern Economy in Historic Perspective* (Stanford University Press, Stanford, California), pp. 115–65.

Pinstrup-Andersen, P. (1988) *Food Subsidies in LDCs: Costs, Benefits, and the Policy Options.* Published for the IFPRI (The Johns Hopkins University Press, Baltimore and London).

Raj, K. N. (1983) 'Agricultural Growth in China and India – Role of Price and Non-price Factors', *Economic and Political Weekly*, vol. XVIII, no. 3, January 15.

Rayner A. J. (1974) 'Regional price differentiation policies for milk in England and Wales: An empirical evaluation using a spatial equilibrium model', unpublished PhD thesis, Department of Agricultural Economics, University of Manchester, England.

Research and Development Centre of the State Council (1989) *Survival, Reforms and Development* (Zhangavan Publishing House, China).

Research Centre of Economic Policy, MOA (1991) *Rural China: A Review of Policy Research*, vol. II (Reform Publishing House, Beijing).

Riskin C. (1987) *China's Political Economy – The Quest For Development Since 1949* (Oxford University Press, New York).

Samuelson, P. A. (1948) *Foundation of Economics Analysis* (Harvard University Press, Cambridge, MA).

Samuelson, P. A. (1952) 'Spatial Price Equilibrium and Linear Programming', *American Economics Review*, vol. 42, pp. 283–303.

Scobie, G. M. (1983) *Food Subsidies in Egypt: Their Impact on Foreign Exchange and Trade*, IFPRI, Research Paper no. 40.

Sicular, Terry (1993) 'The quest for sustained growth in Chinese agriculture', in A. J. Rayner and D. Colman (eds), *Current Issues in Agricultural Economics* (Macmillan Press, London).

SSB (1960) *Ten Great Years* (Beijing: Foreign Language Press).

SSB (1980b) *Main Indicators, Development of the National Economy of the People's Republic of China (1949–79)* (Beijing: Statistical Publishing House).

SSB (1981a) *Chinese Statistical Yearbook 1981*, Overseas Chinese Language Edition (Economic Reporter Publishing House, Hong Kong).

SSB (1981b) 'Communique on Fulfilment of China's 1980 National Economic Plan', *Beijing Review*, vol. 20, pp. 17–20.

SSB (1981c) *Selected Economic Statistical Materials 1949–79* (Statistical Pubishing House, Beijing).

SSB (1982a) 'Report on the Results of the 1981 Plan Implementation', *JMJP*, April 30, pp. 2–4.

SSB *CAYB*, various issues, 1985 to 1991 (Statistical Publishing House, Beijing).

SSB *CSYB*, various issues, 1982 to 1993 (Statistical Publishing House, Beijing).

Takayama, J. and G. G. Judge (1971) *Spatial and Temporal Price Allocation Model* (North Holland, Amsterdam).

Tang, A. M (1984) 'An Analytical and Empirical Investigation of Agriculture in Mainland China, 1952–80', *Taipei, Taiwan: Chung-Hua Institution of Economic Research*, vol. XVIII, pp. 230–251.

Tang, A. M. and B. Stone (1980) *Food Production in the People's Republic of China*, IFPRI, Research Report 15.

Timmer, C. P. and J. R. Jones (1986) 'China: An Enigma in the World Grain Trade', in J. R. Jones (ed.), *East-west Agricultural Trade* (Westview Press, Colorado, USA), pp. 153–180.

Van De Panne, C. and A. Whinston (1964) 'The Simplex and the Dual Method for Quadratic Programming', *Operations Research Quarterly*, vol. 15, pp. 355–88.

Walker, K. R. (1988) 'Trends in Crop Production 1978–86', *China Quarterly*, December 1988, pp. 592–620.

Wei, Shuangfeng (1987) *Studies on the Current Grain Problems and Solutions in Bolo County*, unpublished discussion paper issued by the Investigation Team of the South China Agricultural University for the Grain Problems of Bolo County, Guangdong Province, China.

Weinschenck, G.; W. Henricksmeyer and F. Aldinger (1969) 'The Theory of Spatial Equilibrium and Optimal Location in Agriculture: A Survey', *Review of Marketing and Agricultural Economics*, 37.

Wolfe, P. M. (1959) 'The Simplex Method for Quadratic Programming', *Econometrica*, vol. 27, pp. 382–98.

Wong Chong Ming (1978) 'A Model for Evaluating the Effects of Thailand Government Taxation of Rice Exports on Trade and Welfare', *American Journal of Agricultural Economics*, vol. 60(1), pp. 65–71.

World Bank (1991) *China: Managing an Agricultural Transformation – I (part I – grain sector review)*, working papers, Volume I: Working papers 1–3, China Department, Agricultural Operations Division, Asia Regional Office.

Wu Shuo (1988) 'Analysis and Policies on Short-term and Medium-term Grain Problems', *Problems of Agricultural Economy*, no. 11.

Yan Rui-zhen (1985) 'Economic Reform in Rural China', unpublished document, Dept of Agricultural Economics, University of Manchester.

Yang Chien-pai and Li Hsueh-tseng (1980) 'On the Historical Experience of the Relations between Agriculture, Light Industry, and Heavy Industry in China', CKSHKH, 3:19–44 (English version in Social Sciences in China 2:182–212)

Yao, S. (1987) *Chinese Agricultural Policies and the Grain Problems*, Unpublished MA dissertation, Department of Agricultural Economics, University of Manchester.

Yao, S. (1988) 'Chinese agricultural reforms and prospects', paper presented to the UK Agricultural Economics Society Annual Conference held in Manchester, April.

Yao, S. (1989a) 'From the Point of View of Comprehensive Economics, Re-assess the Role of Agriculture in the Chinese Economic Development', in Wei Shuangfeng (ed.) *Selective Papers on Comprehensive Economics* (Ghuangdong Higher Education Publishing House, Guangzhou, China).

Yao, S. (1989b) 'Policy Modelling for the Chinese Grain Sector within a Framework of Quadratic Spatial Equilibrium', *Indian Journal of Quantitative Economics*, vol. 5, no. 2, pp. 49–68.

Yao, S. (1989c) 'Impact of Chinese Agricultural Policies upon the Patterns of Specialization in Grain Production: An Application of a Quadratic Spatial Equilibrium Model in the Chinese Grain Sector', unpublished Ph.D thesis, Department of Agricultural Economics, University of Manchester, England.

Yao S. and D. Colman (1990) 'Chinese Agricultural Policy and the Grain Problem', *Oxford Agrarian Studies*, vol. 18, no. 1, pp. 23–34.

Yao, S. and R. Hay (1991) 'Food Market Liberalization: History and Prospects', *Oxford Agrarian Studies*, vol. 19, no. 2, pp. 73–90.

Yao S. and Xie Jianmin (1991) 'Determinants of Grain Productivity in China: A Time-series and Cross-sectional Analysis', Paper presented in the international conference on Rural Development in the Open and Coastal Areas of China, held 12–17 December 1991, Zhongshan, Guangdong, China.

Yu Kuo-yao (1980) 'Further Understanding of Problems in China's Agricultural Development Policy', *HC* 5:28–30.

Yueh Wei (1981) 'Questions Concerning the National Income', *CCYC*, 2:46–52.

Zhao Zhi-yang (1988) 'Report at the Meeting of the 3rd Plenum of the 13th Party Congress of the Communist Party of China', *People's Daily*, 28 October, pp. 1–2.

Zheng Zhong (1988) 'New China's Grain Production in Retrospect', *Problems of Agricultural Economy*, no. 2, pp. 3–6.

Index

Index